ENCOUNTERING THE NEW WORLD
1493–1800

The publication of this catalogue has been made possible by a major grant

from FIDELITY INVESTMENTS through the FIDELITY FOUNDATION,

and by the support of the NATIONAL ENDOWMENT FOR THE HUMANITIES,

an independent agency of the federal government.

Frutex Lauri folio pendulo, fructu
tricocco Semine nigro Splendente.

Red Wood.

Psitticus Paradisis.
The Parrot of Paradise

ENCOUNTERING THE NEW WORLD 1493 TO 1800

Catalogue of an exhibition by SUSAN DANFORTH

With an introductory essay by WILLIAM H. MCNEILL

THE JOHN CARTER BROWN LIBRARY

PROVIDENCE, RHODE ISLAND

1991

ISBN 0-916617-37-8

Venues for the public presentation of
Encountering the New World, 1493 to 1800

February 16 to April 9, 1988
IBM Gallery of Science and Art (New York City)

January 11 to March 8, 1992
Amon Carter Museum (Fort Worth, Texas)

April 12 to May 31, 1992
The Tennessee State Museum (Nashville)

June 28 to October 25, 1992
Natural History Museum of Los Angeles County

Winter – Spring 1993
Museum of Art, Rhode Island School of Design

Tour organized by the Trust for Museum Exhibitions
Washington, D.C.

CHRISTOPHER
COLUMBUS

QUINCENTENARY
JUBILEE

OFFICIAL QUINCENTENARY PROJECT

FRONTISPIECE: See figure 7.

CONTENTS

PREFACE

THE COMMEMORATION now beginning in the United States of the fifth centenary of Columbus's historic voyage to the West has evoked startling controversy. Some descendants of the native inhabitants of the Americas in 1492 would prefer that the anniversary on October 12th be a day of lamentation rather than a prideful celebration. Others, whose primary concern is the protection of the global environment, see in the European conquest of the Americas the beginning of a tendency, made possible by European technological and industrial prowess, toward horrific and heedless despoliation of the beauty and resources of the earth.

Who can doubt that the incredible European expansion across the entire globe that began, in Columbus's lifetime, with the Portuguese explorations of the west coast of Africa and continued most spectacularly with the Spanish-sponsored voyage of discovery that led to landfall in the Americas, unleashed gigantic and unprecedented forces that have transformed the world? But such being the case, let us by all means, let us urgently, mark the day, the year, the centenary, and the quincentenary. Let us study and reflect upon the multiform meanings of the unprecedented encounter, celebrating, lamenting, or condemning as anyone may wish, but wringing from these events and their consequences every drop of understanding that we can. To counsel that October 12, 1992, or October 12 in any year, be condemned to the oblivion of non-remembrance would be ostrich-like indeed.

The John Carter Brown Library, a center for advanced research and one of the world's great repositories of printed materials contemporary with the age of European discovery and exploration and with the colonization of the New World by the Spanish, Portuguese, French, Dutch, and English, is dedicated to the promotion of research and writing pertaining to the great encounter and its colonial aftermath. The Library also hopes to contribute to public enlightenment concerning the history of this fateful 350-year period, from the first tentative Portuguese ventures into the Atlantic to the revolutions for independence in the New World led by such figures as George Washington and Simón Bolívar.

The present exhibition and catalogue is one of a number of different ventures in which the John Carter Brown Library is continually immersed that concern specifically the European reaction to the discovery. In June of 1991 the Library sponsored a four-day, international scholarly conference on the subject of "America in European Consciousness: The Intellectual Consequences of the Discovery of the New World, 1493–1750," which was focused on the question of what difference it made to European thought, culture, and sensibilities, from the creation of art to the writing of theology, that the Americas had been discovered. The Library is also engaged in the publication of a six-volume chronological guide to European books about the Americas published between 1493 and 1750. Entitled *European Americana*, this massive work, some twenty years in preparation, will contain about 31,000 entries when it is completed early in 1993. Four volumes of the six have already been published.

The traveling exhibition "Encountering the New World" and this accompanying catalogue are somewhat different from both of the above-mentioned projects. The present exhibition and catalogue, unlike the projects mentioned above, concentrate, so far as possible, on the subject of what Europeans could know and did know of the New World *visually* before ca. 1800, that is, well before the era of photography or even of mass produced lithography. The replication of visual images in this age depended upon the woodcut and the engraved copper plate, and the curator of the exhibition, Ms. Susan Danforth, took it upon herself to scour the John Carter Brown Library collection to answer the question of what was actually available to Europeans in the way of pictures of peoples, plants and animals, urban settlements, landscapes and landmasses in the Americas. The analysis of such images, of almost any one of them, could be the subject of many dissertations and could result in a very big and learned work. But Ms. Danforth's goal here, and the Library's, was not so much to analyze as to collect and present a broad range of visual materials, thereby increasing both scholarly and popular awareness of what existed at the time and of its character.

Whereas the exhibition itself, with nearly two hundred items on display, is organized in the form of a continuous narrative, this catalogue, being a radical selection from the whole, is more of an album or a sample extracted from the exhibition. It gives the flavor of the exhibition, but it does not attempt to tell the story. On the other hand, we do intend for the catalogue to be a complete record of what is in the exhibition, and every item is therefore recorded in a short-title list at the end, with an identifying photo.

As always, the production of a work such as this incurs debts. We wish first to thank Ms. Barbara Walzer, who assisted Ms. Danforth as a picture researcher when the exhibition was first planned in 1986 and 1987. The exhibition opened initially at the IBM Gallery of Science and Art in New York City in February 1988. This early association with IBM was enormously beneficial to the development of the exhibition. The experienced curatorial and design staff at the IBM Gallery helped the JCB staff to refine the concept of the exhibition and its presentation. We wish to acknowledge also the good work of Mr. Richard Hurley, the Library's staff photographer, who took all of the photos used in this catalogue.

We are indebted, too, to Professor William H. McNeill, the distinguished scholar of world history, for agreeing to write an introduction to this catalogue.

Funding for the traveling exhibition was provided in part by a grant from the National Endowment for the Humanities, which included support for several of the Library's Quincentenary projects. This NEII grant required matching money, which was provided by the Ahmanson Foundation, the Florence Gould Foundation, and Fidelity Investments through the Fidelity Foundation. To all of these generous donors the Library is deeply grateful. We wish particularly to acknowledge here the donation from the Fidelity Foundation, which provided support specifically for the cost of manufacturing this catalogue.

The John Carter Brown Library does not ordinarily allow any of its materials to leave the Library's premises. These precious books, maps, prints, and other documents from the past, a good number of which are known to exist in only a few copies, or which may even be unique survivors, cannot lightly be put at risk. However, the 500th anniversary of Columbus's earth-uniting voyage seemed to the Library's Board of Governors to be an occasion of such magnitude that an exception could be made. It is our hope that those who see the exhibition will be helped by it to imagine the experience of Europeans in the sixteenth, seventeenth, and eighteenth centuries as the reality of the new, and surprising, American continents invaded their thinking and compelled rethinking.

NORMAN FIERING
Director and Librarian
John Carter Brown Library

To the Reader

The present work has two major sections. The first consists of illustrations of a selection of about 76 items extracted by the curator, Susan Danforth, from the full exhibition, which has nearly 200 items in all. This first section, then, with figure numbers, is intended to be a kind of album that provides a sampling, often in color, of the full range of visual material in the show. The second section, separately organized at the end of the catalogue, is a comprehensive list of all of the items in the exhibition, accompanied by small, identifying, black and white illustrations. This section provides a full record, for future reference, of what is in the exhibition.

INTRODUCTION

The Legacy of Columbus, or How by Crossing the Oceans He Shaped the Modern World

COLUMBUS'S FAMOUS VOYAGE, whose 500th anniversary we will celebrate in 1992, triggered a process of interaction across the oceanic spaces of the earth which is still in train. He made the world one as never before; and that is why his voyage was, and still is, so important. Others had crossed the Atlantic before him—Leif Erickson for certain, along with others whom we do not know by name. A few Vikings actually settled in Newfoundland, and humble fishermen from Europe had probably been catching cod off that island for a generation or more when Columbus set sail, though they kept the Grand Bank's marvellous fishing so secret that we cannot be sure. But since none of the earlier crossings led to continuous linkages between the Old World and the New, they are merely historical curiosities, and not the epoch-making event that Columbus's voyage turned out to be.

Why did the recently discovered Viking settlement in Newfoundland die out without making any difference to American or European life, and why did Columbus's report of his first voyage trigger such fateful, world-changing consequences? The answer lies in the different scale of follow-up the two enterprises provoked.

When Lief Erikson sighted North America, soon after the year 1000, his discoveries circulated only by word of mouth among a small population of Scandinavian seafarers. For a brief time a handful of them actually tried pioneering in Newfoundland, but most who heard the news were busy with shorter, safer voyaging, and saw no advantage in crossing wide and stormy seas to settle far away. Even when, centuries later, written sagas put a report of Leif's discovery into more permanent form, only a few literate Icelanders could understand them.

No one thought of doing anything about it by then, for Leif's home base in Greenland had fallen on hard times and was about to disappear, owing, largely, to a change of climate that made a European style of life in Greenland all but impossible.

By contrast, in 1493, when Columbus returned from his first voyage, his report was printed at once and translated into several languages. His news attracted wide attention. Important people quickly decided to organize new and better equipped voyages across the Ocean Sea. Queen Isabella, who had backed the first voyage, was no longer alone in favoring Columbus. Instead, the full resources of the Spanish crown were put behind the overseas adventure. Officials of the government confidently called on Genoese bankers to finance further voyages, expecting spectacular profits from trade and conquest in the Indies. Even when Columbus's initial hopes proved false, and the fabled wealth of the Indies turned out to lie across another and far wider ocean, the Pacific, the Spanish government continued to despatch ships, men, animals, plants, tools, and weapons across the Atlantic. The adventurers who thus arrived in the New World soon managed to seize spectacular amounts of gold and silver, first in Mexico (1521) and then in Peru (1533). News of their feats aroused dreams of sudden wealth throughout Europe, and persuaded sober men of affairs to keep on sending men and supplies to America, even though most ventures failed to repay the cost of fitting them out. But the few that did succeed brought such bonanzas that ships kept coming. As a result, Columbus's voyage triggered a continual encounter that has lasted down to the present.

Windfalls of gold and silver soon gave way to mining and to plantation agriculture as the main forms of European enterprise in the New World; but agricultural settlements, which recreated European styles of life across the Atlantic as fully as local conditions permitted, also eventually established themselves. Little by little the drastic initial differences between the two sides of the

Atlantic Ocean diminished, without ever disappearing entirely. That, in a nutshell, is the history of the Western world across the past five hundred years.

Columbus started it all; but there was nothing so very remarkable about his achievement. If he had not crossed the Atlantic in 1492 some other European sailor would surely have done so very soon afterwards. Columbus was not the only ship captain who understood the wind pattern of the Atlantic which made sailing west and east at will easy to arrange. By sailing south to the trade wind zone, as Columbus did, a following wind carried sailing ships safely westward; and by heading north into the zone of prevailing westerlies, as Columbus also did, the return voyage could also count on generally favorable, though sometimes stormy, winds. Under the circumstances, and given the seafaring skills that Europeans had developed by 1492, it is safe to say that the first trans-Atlantic voyage was only a matter of time. Proof is the fact that Pedro Cabral, a Portuguese captain heading for India along the route Vasca da Gama pioneered in 1499, blundered into the coast of Brazil in 1500 simply by taking a more westerly course than da Gama had done.

The really surprising aspect of the European discovery of America, therefore, was not the voyage itself but the relentless way it was followed up. By 1493, Europe's political, economic and intellectual organization was ready and able to focus relatively vast human and material resources on ventures overseas. They paid off so handsomely that a feedback loop came into action whereby expansion in the Americas sustained and reinforced Europe's capability for further overseas enterprises. It was, or swiftly became, a self-sustaining process. We are heirs of that post-Columbian process, still struggling to understand what it was, and is, and how it will continue into the future.

For a closer look at Columbus's legacy it is convenient to separate 1) ecological consequences, 2) political and social phenomena, 3) intellectual and cultural changes, even though the post-Columbian encounter was a single affair, embracing all aspects of life all the time. But our minds cannot cope with undifferentiated complexity. Words clarify and separate what in reality is indistinct and interconnected. This is the secret of our mastery of the natural world, since by acting on a simplified image of reality we can make it more to our liking. Thinking about ourselves and how we got to be the way we are requires the same simplification, even though human behavior never fits tidily into this sort of artificial compartmentation.

First, then, ecology. Historians are only beginning to realize how drastic the ecological collision between the Old World and the New turned out to be. Initially, flora and fauna as well as human races and cultures of the Americas were entirely different from those on the other side of the ocean. Generally speaking, the Eurasian and African ecosystems were more highly developed than those of the Americas. That was because the larger area of the Old World allowed more chances for evolutionary alteration of existing life forms, favoring the more successful variants; and a tighter system of communication within Eurasia, developed by humans across the centuries of recorded history, meant that particularly successful forms of life had spread very widely, coming close to their geographic limits. Crops and domesticated animals had been carried far and wide by deliberate human action. In addition, other organisms travelled with ships and caravans, including weeds and pests that were inimical to human purposes, and disease germs that endangered human life. Nevertheless, Eurasian-African life forms were more or less in balance, having adapted across the centuries to survival in each other's presence. The same was true of the Americas, but the web of life in the New World had never faced competition as intense as that which prevailed in the Old and was therefore unable to hold its own against a great many of the invaders that arrived from the other side of the ocean.

From 1492 onwards a great many Eurasian and African organisms were therefore poised and ready to establish themselves in the Americas. Doubtless, many failed to make lodgement in the New World and some took a long time to get across the ocean barrier. But those that did often found conditions very hospitable and flourished exuberantly because native fauna and flora were unable to compete with them on even terms. The resulting catastrophe to native American forms of life was enormous, and, as Alfred Crosby explained in his important book, *Ecological Imperialism* (New York, 1986) the invaders tended to support one another, simply because they were already adjusted to living together. Invasive grasses that nourished horses and cattle, for instance, withstood the trampling of their hooves far better than native American grasses could do; and this favored their propagation as soon as vast herds of animals from the Old World began to run wild in the Americas.

From a human viewpoint, by far the most important organisms that Europeans brought to the Americas were disease viruses. Smallpox, measles, influenza, whooping cough, and other such infections killed a large number of infants and children in Europe, where they had been endemic for centuries. But high birth rates allowed human numbers to maintain themselves, or even to grow, despite such losses. In other words, a more or less stable balance between Europe's endemic disease organisms and their human hosts had been achieved across the centuries. In Africa, an array of tropical diseases, most notably malaria and yellow fever, supplemented the roster of those established in Europe. But African populations had become accustomed to tropical diseases and developed resistances that other peoples lacked.

By contrast, on first exposure the native peoples of the New World were utterly vulnerable to Old World diseases. Lacking both inherited and acquired immunities, old and young were equally at risk. Endemic childhood diseases of Europe therefore became killer epidemics for adults as well as for children in America. The first to show its power was smallpox, which arrived in Mexico with the expedition that reinforced Cortes in 1520. Its devastating consequences made his final victory possible. In subsequent years, each new disease renewed the catastrophic initial impact of smallpox, so that a long succession of lethal epidemics tore down the fabric of Indian society and culture. Native priests and rulers did not know how to cope with infections that made nearly everyone seriously sick and killed up to a quarter of the entire population in a matter of days. Drastic demoralization ensued.

The old gods, together with those who had served them, were totally discredited. The lopsidedness of the disease onslaught relentlessly drove the lesson home, for Spaniards, having met such infections in childhood, were usually untouched while native Americans died like flies. The only sensible response in such a situation was to come to terms with the Spaniards' God, who protected them so effectually. As a result, mass conversion to Christianity met with no effective opposition from discredited pagan priesthoods or frightened chieftains. Native converts often became more pious than the Spaniards, even though (or perhaps because) they continued to suffer disproportionately from imported diseases until about 1650. After that time a new disease equilibrium began to emerge in Mexico as the major viral diseases of the Old World became endemic there just as in Spain.

Before that happened, during the first hundred and thirty years of contact, the ravages of disease reduced the pre-Columbian population of Mexico from something

between 15 and 25 millions to a mere 1.6 million! Peru suffered similarly catastrophic depopulation; and other regions of the Americas all suffered profound disruption whenever contact with outsiders brought the deadly viruses into range. Many communities were wiped out entirely. Everywhere, resistance to European encroachment was weakened. Remember the school book story of how Squanto showed the Pilgrim Fathers the way to plant corn in fields that just happened to be ready and waiting for them! A lethal epidemic (perhaps originating in Jamestown) had killed off most of Squanto's people shortly before the Mayflower arrived in 1620. As a result, the few, shattered survivors could not oppose the Pilgrims even if they had wanted to. Squanto's story is therefore aptly symptomatic of how epidemics cleared the way for the advancing whites. Obviously, without the Indians' vulnerability to imported diseases the whole history of the Americas and relations between indigenous and immigrant human populations would have been fundamentally different.

Why did the indigenous peoples of the New World not have diseases of their own with which to retaliate? Perhaps they did infect Columbus's sailors with syphilis, which first came to the attention of Europeans in 1494 when an epidemic of the unfamiliar disease broke out in an army besieging Naples. But syphilis was a trifling counterpart to the long list of killer epidemics that travelled the other way. This is proven by the fact that it made no perceptible difference to the demography of the Old World even though it attracted a great deal of attention.

This remarkable lopsidedness arose because ancestors of the original human inhabitants of America had travelled across the Bering straits in small bands, and therefore arrived without diseases that can be sustained only when there is a large population of vulnerable newborns to keep the chain of infection going. Even after Amerindians settled down to agriculture and their populations became large enough to sustain such crowd diseases, they remained immune because they had no domesticable herd animals that harbored such infections naturally. By way of contrast, across thousands of years a number of transfers from animal herds to human hosts did take place in the Old World, exposing Europeans to the viral diseases which wreaked such enormous damage upon the American Indians. This is a very telling example of how the Old World's more complex and highly evolved ecosystems gave organisms accustomed to survival there an enormous advantage in competition with those which had escaped comparably intense pressures.

The resulting ecological imbalance went a long way to define the course of the encounter between the Old and New Worlds. Yet at the time no one understood what was happening, and the handful of ruthless adventurers who conquered and then ruled the vast empires of Mexico and Peru did not wonder why it was possible for them to prevail so easily, or subsequently to spread their religion, their language, and their literate culture far and wide. They knew it was God's will, and for centuries this simple explanation diverted attention from the cloud of disease germs and other organisms they brought with them to the Americas. But historians today do not think that God plays favorites, and are coming to believe that horses and (very inefficient) guns were not enough to assure Spanish victory. Instead, it was ecology that made the difference, upsetting pre-existing biological and cultural balances completely and allowing the human intruders to prevail because the ecosystem of which they were a part was prevailing as well.

Yet, as in any ecological encounter some American organisms held their own easily enough and soon found new niches in the Old World as well. Aside from syphilis, whose American origin is not certain, American food crops constitute by far the most notable example of this reverse flow. Potatoes, maize, sweet potatoes, tomatoes, peanuts, peppers, cacao competed quite successfully in the Americas with crops the Europeans brought with them; and indeed, white pioneers prospered in North America only after learning to rely on maize as their principal crop.

Maize and potatoes also spread widely in Europe, Africa, and Asia. Their fundamental attraction for Old World farmers was that in suitable soils and climate these crops yielded far more calories per acre than any of the Old World grains, with the exception of rice. Accordingly, when population growth made it worth the extra effort required for hoeing potato and maize fields, these two crops entered Old World crop rotations, and in some important regions actually displaced older staples. This was true, for instance, in southern Africa, where maize superseded millet as the basic food; and in northern Europe, where potatoes became more important than rye, which was the only grain that would ripen in the cool, wet climates of the north European plain.

By deliberate borrowing from the Americas, Old World peoples were thus able to benefit very substantially, and escaped nearly all of the ecological disasters that the opening of regular contact across the ocean brought to the plant, animal, and human populations of the Americas. To be sure, Europeans did have a few brushes with disaster, most notably when a fungus, newly imported from Peru, destroyed the potato crop in Ireland and in much of Europe in 1845 and 1846. Up to one million persons died in Ireland from the ensuing famine; but this catastrophe, severe though it was, pales in comparison with the repeated disasters that struck the native populations of the Americas after 1520.

So much for the biological disturbances provoked by the opening of contacts across the ocean. Obviously, ecological upheaval constituted an essential background for political, social, and intellectual sides of the European venture overseas, since it gave Europeans and Africans an overwhelming advantage in their encounters with native peoples. What they made of that advantage concerns us next.

Turning first to politics and society, we in the United States have long cherished a very lopsided view of how settlers from Europe responded to American circumstances. The ravages of lethal disease meant that very soon after the initial European lodgements in the New World, manpower shortages became general. Europeans knew very well how to farm, how to dig ore and refine precious metals, and how to organize a great variety of other activities that were essential for civilized life as practiced in Europe. But few white people actually crossed the ocean since that was always a costly undertaking. Europe's poor could not afford passage, unless someone paid their way. Who then would do the dirty work required to maintain the fabric of civilized society?

In Europe, birth and death rates maintained an abundant supply of poor laborers who took on nasty jobs simply to get food enough to keep themselves and their families alive. But in the Americas, nothing of the kind was possible. Catastrophic death rates among the native population meant that manpower needed for village routines was already short. No one would willingly leave home to work at new tasks under alien masters. And the relatively small number of Europeans who were on the scene felt exactly the same. Why work at nasty jobs when economic necessity did not require it? And economic necessity did not compel immigrants to work for others since open frontiers, recently denuded of their Indian inhabitants, beckoned the adventurous.

This circumstance created drastic alternatives for government and society in the New World. If everyone were at liberty to do what he wished, the social pyramid familiar in Europe (and in civilized pre-Columbian societies of the Americas as well) would simply collapse. No one would have to serve social superiors. Large-scale enterprises, like deep rock mining, that required the coordinated activity of many different persons would become impossible. But widely-dispersed, free, and more or less equal families could make a living by scattering out across the countryside, relying on hunting and gathering to supplement subsistence agriculture. Many pioneers lived in such a fashion, needing only a few items that they could not produce for themselves — though guns, knives, and axes were always important and could only be acquired through trade links that reached back to specialized artisans who lived in cities.

The problem was, however, that frontiersmen and pioneers, by leaving cities with their occupational specialization and social hierarchy behind, soon descended into a state of neo-barbarism. They simply could not maintain most of the skills of civilized life and had only slender connections with those who did. From the point of view of European rulers, these neo-barbarous, free and equal pioneers made very unsatisfactory subjects, since they paid no taxes and (usually) performed no services for the government, although armed raiding parties recruited among frontiersmen were sometimes effective in colonial wars of the eighteenth century.

The alternative that appealed to European authorities therefore was to maintain social subordination in the Americas by instituting some sort of legal compulsion. Slavery was the most obvious resort; but because Spanish jurists decided that enslaving the native peoples of America was illegal, the alternative of peonage was invented to compel Indians to work at tasks the Spaniards wanted them to perform. The Portuguese in Brazil had no such qualms; but in fact, enslaved Indians usually died of disease soon after being captured and carried away from their native villages. Indian slavery therefore never became important in Brazil, and the same was true in the English and French settlements of North America for exactly the same reason.

Africans were far more resistant to the diseases that ravaged American Indians so cruelly, and when this became clear, an organized slave trade arose that carried millions of Africans across the Atlantic. They worked for white masters producing goods for the distant markets of Europe. Sugar became the most important product of enslaved labor in the Americas; but cotton and other commodities, like brazilwood and indigo, were also despatched to Europe on the strength of the compulsory labor of African slaves and their descendants.

Peonage had a similar effect. Spanish officials and landowners in Mexico and Peru acquired or usurped the legal right to control Indian labor, and used it to work technically very efficient silver mines, to erect churches and governmental buildings, and to manage ranches and agricultural estates. Spanish authorities were therefore able to create a similacrum of European cities in the New World with amazing rapidity when their empire was new; but as the supply of labor diminished, maintaining the full panoply of urban life became difficult. Nearly self-sufficient estates, where ranching was more important than tillage, multiplied, and cities withered proportionately. Nevertheless, the command structure of colonial society persisted, and with it a ruling class of Spanish officials, clergy, and landowners who kept in touch with the homeland and with European culture in a way that rude backwoodsmen and pioneers were quite unable to do.

In Brazil and the Caribbean, where sugar was king, slave plantations fell far short of the aristocratic and official elegance of Mexico City and Lima. Nevertheless, in Brazil and the sugar islands of the Caribbean, as also in Virginia and the Carolinas where tobacco and cotton were the important crops, slave labor supported a class of plantation owners and managers who sold their products on the European market and shared the culture of their ancestral homelands almost as fully as did the officials and clergy of the Spanish empire.

The New World thus became the home of social and political extremes. In some parts, neo-barbarism and frontier equality prevailed; and historians of the United States, ever since Frederick Jackson Turner's famous speech at the Columbian Exposition of 1893 on "The Influence of the Frontier in American History," have emphasized the virtues and formative significance of that version of American society. Yet slavery also flourished mightily on the frontier in the southern part of the United States, especially after the invention of the cotton gin in 1794 made it possible for slave labor to produce large amounts of fiber for the burgeoning cotton mills of England.

Indentures whereby immigrants repaid the cost of passage across the ocean by agreeing to work at the bidding of a master and without pay for as much as seven years was another form of compulsory labor that played a considerable role in maintaining civilized social complexity in the English colonies of North America, especially in the early decades when Black slaves were not yet very numerous. Most of those who came to the English colonies from Europe before 1800 came as indentured servants. They may have come in hope of winning freedom and equality; but their initial experience was harsher than was normal in England, where indentured labor was the fate only of youthful apprentices.

Thus, if one takes the colonial era in all parts of the New World into consideration, our cherished national image of America as a blessed haven of freedom and equality is somewhat askew. Neo-barbarism existed, but was far outweighed by the achievements of legally enforced social hierarchies. The Puritan polity and society of New England, which patriotic United States history treats as typical of the New World, was in fact exceptional and not so very free. Ministers and elected officials enforced a severe discipline in matters of deportment and religion. A vigorous merchant class soon established itself in Boston and other sea ports, trading overseas with the Caribbean and with Europe. They paid higher wages than were common in Europe, but demanded and got extra effort in return. Social subordination was therefore real even though the gap between rulers and ruled was far less in New England than in other parts of the New World.

A balanced view of American society and government in the first three post-Columbian centuries therefore must recognize the prevalence of compulsory labor in the New World. That was the only way colonists could produce goods for export to Europe and maintain the fabric of civilized society and government, with all the differentiation of skills and social roles that civilization required. Yet a society built upon compulsory labor was different from one in which the balance of supply and demand in the labor market assured social subordination and differentiation of roles. And, of course, the egalitarianism that prevailed among remote and ungovernable frontiersmen was also different from the social hierarchy of Europe, but in the opposite direction.

In other words, American society and government ran to extremes as compared to Western Europe. Exactly the same was true on the eastward frontier of European civilization, and for the same reasons. For in Russia, as in the Americas, beginning about 1550, a neo-barbarous, subsistence style of egalitarianism in Siberia and along the southern frontier competed with the compulsory labor of newly enserfed peasants, who obeyed their social superiors by producing grain and other commodities for export. And, as in the New World, until population growth made the abolition of serfdom in the nineteenth century seem practicable to Russia's rulers, persistent manpower shortages meant that compulsory labor affected more people and played a far larger role in government, trade, and war than its opposite.

Despite this resemblance, American society differed from that of Europe's eastern frontier in one important respect. From the start, intermingling of races and cultures prevailed in America far more obviously than in the Old World, where racial and cultural intermingling was far older and lines of demarcation were therefore blurred. In Russia, for instance, linguistic and religious diversity along the frontier was very great, but the various peoples who were eventually swept into the Russian empire all counted as white until Tsarist officials collided with the Chinese and divided sovereignty over the Far East with them after 1689. Nearer the heart of European civilization, west European nations were comparatively homogeneous, thanks to centuries of immunity from foreign conquest and the educational efforts of Church and state.

The nearest Old World analogue to the racial-cultural demarcations that divided American society existed in the heartlands of the Islamic world, where a ruling class, harking back to Turkish origins, lorded it over other ethnic groups. In Middle Eastern cities diverse peoples assorted themselves into distinct quarters and followed various trades, each of which tended to be monopolized by a particular ethnic group. Differences of wealth, prestige and of religion established an order of precedence among the various ethnic elements. Each of them preserved its own customs, culture and demeanor, adapting to the social peck order by deference to superiors and demanding deference from inferiors.

In the Americas, a somewhat similar polyethnic social hierarchy became, as it were, indelible because differences of physical appearance came to be associated with social status and occupation. Racial differences and social differences tended to coincide, thereby reinforcing one another and making it difficult or impossible for individuals to change status on the strength of personal skills and accomplishment. Whites were on top; Blacks at the bottom; and various race mixtures of white, red and black occupied intermediate social positions. Racial and cultural differences therefore made the legal reinforcement of social hierarchy far more rigid than it might otherwise have been, simply because every individual carried outward markers of the social status and identity conventionally assigned to his race. Upward mobility under such circumstances faced obstacles far greater than prevailed in racially more homogeneous societies.

Influence, as always, ran both ways. Thus, the rapid incorporation of the Americas into the European political and economic system also reinforced some forms of social hierarchy in the Old World. In particular, emerging national states found it possible to exercise sovereignty at a distance, and thereby took on new functions and acquired new revenues. The Spanish government was by far the most successful in this respect; but France, Britain, Holland, and Portugal did their best to match the achievements of Spanish officials and actually overtook them in course of the eighteenth century by favoring and protecting merchants' overseas ventures more effectively than aristocratic Spanish officials were willing to do. When reinvested in ships and merchandise, the wealth derived from transoceanic trade resulted in rapid capital accumulation; and swelling amounts of private capital formed a tax base that paid for the newly professionalized navies and armies that allowed Europeans to exert sovereign power across enormous distances. Such changes might have come to Europe after 1492 even if the Americas had never existed; but in fact they did exist, and exploitation of American resources on an ever increasing scale played a prominent part in strengthening the national states that succeeded in planting colonies across the Atlantic.

Bankers and capitalists constituted another group that increased its power by dealings with the Americas. Their relation with state officials was ambiguous, for they needed armed protection, which governments could give, but simultaneously detested taxes, which governments imposed. Capital flowed across political frontiers toward countries where protection was effective and taxes were minimal. The sudden rise of Holland in the late sixteenth century was a clear instance of this phenomenon; and London's rise to commercial and financial primacy after 1660 resulted from the same factors. In effect, a rather sticky market in protection costs operated among the European states, so that officials who taxed too much and provided too little protection in return soon found their revenues shrinking because capital and trade moved elsewhere.

This happened to Spain after about 1550, impoverishing a country whose imperial responsibilities soon began to outrun its resources. By the eighteenth century, what defended the Spanish empire in the Americas was no longer the military power of the home government but the power of yellow fever to destroy expeditionary forces from Europe. What happened was that after about 1650, when yellow fever first crossed the Atlantic in slave ships, one of the most powerful disease barriers that protected Africa from outside encroachment began to operate in the New World as well. Once yellow fever became established an an endemic infection throughout the lowland tropics of the Americas (by about 1690), its devastating effect on outsiders was exactly the same as in Africa, and protected Spanish possessions in the Americas from successful assault long after the Spanish navy had ceased to be able to do so.

As colonial populations grew, American landscapes gradually filled up with human inhabitants again. Demographic recovery among Indians and mestizos became definite in Mexico and Peru after about 1650. In more remote parts, Indian populations continued to decay, and in the remote Arctic and in Amazon jungles a few isolated communities were still passing through the crisis of exposure to civilized diseases in the second half of the twentieth century. But the main demographic impact of the encounter was absorbed by 1700, with the result that after that date Indian and mestizo populations in the Americas began to increase, slowly at first and then at an accelerating rate as their adaptation to the new disease environment became complete.

Black slaves working on sugar plantations in the Caribbean and Brazil did not reproduce themselves as long as the slave trade continued unchecked, presumably because plantation managers preferred to buy a work force ready grown instead of meeting the costs of childrearing. Climate and diet were also probably less favorable than on the North American mainland, where by the time of the American revolution slave populations had begun to grow almost as fast as the white population was doing. In tropical lands, the arrival of African diseases with the slave ships made it difficult for white settlers to thrive. As a result, descendants of African slaves eventually came to dominate most Caribbean islands and wherever sugar plantations flourished on the mainland.

Descendants of European immigrants multiplied very rapidly in cooler climates; and towards the end of the eighteenth century their numbers were such, and the open frontier had moved sufficiently far inland, that land appreciated in value, and a labor market arose that compelled some people to work for others. In due course, a miniature replica of European society started to grow up along the North American coast as urban skills and occupational specialties were successfully transplanted to emerging cities. The fact that labor remained comparatively expensive encouraged efficiency and ingenuity in its application to routine tasks. This had a tonic effect on American economic efficiency, which in the course of the nineteenth century rivalled and often surpassed anything Europe had achieved.

In time, the rise of the United States to its contemporary world position resulted, and with it a revaluation of the colonial past that filtered out the compulsion of labor that had played so prominent a role in early centuries, and glorified the marginal frontiersmen beyond their just deserts. Yet, as we all know, slavery in the United States was abolished only in 1863 and lasted in Brazil until 1888. Peonage disappeared from Mexico in the course of the nineteenth century, though it lingered in some Andean regions into the twentieth. Moreover, race discrimination persists to the present throughout the Americas, despite legislation to the contrary. This shows that freedom and equality, however noble and attractive as ideals, are only one side of the heritage from colonial times. They were not dominant in New World society in the deeper past; and in spite of political rhetoric, these ideals are still only struggling to prevail among American governments and peoples today.

Yet the aspiration for freedom and equality is enormously attractive to the poor and downtrodden of the earth. For that reason, reconciling actual practice with the public commitments of the United States to freedom and equality will remain difficult. As our country enters into closer connection with a global process of political, economic, and cultural interaction, the peoples in Asia, Africa, and South America will affect the domestic and foreign affairs of the United States far more powerfully then heretofore. Indeed, how to deal with the population explosion in the world's poor countries and with the pressures for migration that result will probably become the central question of politics in the twenty-first century, in Europe as well as in the United States.

This means that the push towards new patterns of polyethnicity will be very strong; and newcomers may not readily achieve equality of rights with older inhabitants, especially if they arrive as illegal immigrants. The long drawn out New World experience of how to cope with racial and cultural pluralism is likely to become influential in Europe, Japan, Australia, and South Africa. But American examples run to extremes, as we saw. Choosing between them will probably become a capital question for the politics of the twenty-first century.

If New World patterns of racial and cultural co-existence do affect Old World politics in time to come, it will be by changing ideas about what is right and pos-

sible. That would then constitute another chapter in the story of how the discovery of America by Europeans affected intellectual and cultural affairs in Europe and the world. Yet this, my third theme, is very difficult to disentangle from the complexity of other influences playing upon human consciousness. Who can say what would have happened without the Americas? And who can tell how much of what did happen was due to the discovery of the Americas and increasing information about the New World?

As an aid in approaching this question let me invoke the words of a fellow historian, John H. Elliott, who once wrote: "In discovering America, Europe discovered itself."[1] This lapidary sentence comes close to the heart of what the discovery of America did for the Old World, it seems to me, for just as I learned English by studying French, seeing language from the outside, so to speak, and discovering that it was not as natural as breathing, so Europeans discovered that their way of doing and thinking was not natural, inevitable, and universal, as they had formerly supposed, and instead had serious gaps in it that had to be repaired.

Since cultural and geographical variability was just as real in the Old World as in the New, why should the discovery of America have made any special difference? The answer to this question, I think, is that Europeans (and all the other peoples of the Old World) had fixed ideas about one another and the distant lands different peoples inhabited. Centuries of contact and conflict had shaped preconceptions that easily prevailed, even when contacts intensified. Thus, Vasco da Gama's famous voyage brought India closer to Europe and vice versa, but did not require either party to alter older ideas about each other in any important respect. The same was true for China and Japan and all the lands between. As for Africa, except in the extreme south, lethal fevers made it impenetrable to outsiders before 1850. But the Americas were both easily penetrable and utterly new. In Columbus's wake, people and goods criss-crossed the Atlantic leaping far greater cultural and ecological barriers than anything that could be experienced within the confines of the Old World. Europeans therefore had to stretch their minds to fit the New World into their inherited notions of reality and truth, and by doing so they opened innumerable further avenues for reassessing received ideas of every sort.

1. John H. Elliott, *The Old World and the New,* (New York, 1970) p. 53.

To be sure, the early venturers from Europe were not careful observers, and had no vocabulary to describe what they saw accurately. But unending encounters provoked a persistent effort to fit the new into familiar frameworks until the weight of too much novelty forced old ideas and preconceptions to give way. As that happened, European culture underwent a fundamental metamorphosis. Old certainties and assumptions could no longer be taken for granted. Truth was no longer to be sought primarily or exclusively in ancient, authoritative texts. Instead, prolonged and repeated experience of the strangeness of America (and elsewhere) compelled thoughtful persons to change their minds about all sorts of things, and to endure far more uncertainty than before.

This distressed many people. A common response to intensified uncertainty was (and still is) to reaffirm important traditional truths with redoubled energy. It is not therefore surprising that the dogmatism of the Reformation and Counter Reformation dominated the intellectual scene in western Europe for more than a century after the discovery of America. But intensified religious convictions did not prevent new and discrepant information from pouring into Europe from the New World and other parts of the globe. To make sense of this data, new and autonomous sciences arose, and in spite of scientists' best efforts to capture eternal truth in words and mathematical formulae, radical new ideas had to be invented from time to time to take account of the new information and improved observations that kept on coming in.

Eventually, after three and a half centuries of sporadic revision of received notions, an evolutionary vision of reality took form. The discovery of America and the ensuing surge of information from the New World accelerated this metamorphosis for sure; and may have made it possible. If so, this intellectual reassessment of the nature of things may well be counted as the greatest, most abstract and universal consequence of Columbus's famous voyage. On the other hand, much of the impetus to intellectual change came from within European society itself, and was facilitated by long-standing discrepancies of Europe's Christian and Classical heritages. Individual genius played a part, too, and so did communication nets that stimulated individuals to think new thoughts. American novelties were only part of the scene in which modernity emerged, and there is no way to test what might have occurred without all the discrepant news from the New World that required learned men of Europe to bethink themselves.

The eventual all-embracing metamorphosis of Europe's intellectual posture was, of course, a result of many lesser revisions of established truths. To begin with, cartography was at the cutting edge. Accurate maps were needed by sailors and by imperial administrators; and map makers already knew how to incorporate scattered new data, including mathematically accurate measurements of latitude, into their portraits of the earth. As a result, the coastal configuration of the New World sprang quickly into being on maps. Initial errors could take on a life of their own as one map maker copied his predecessors; but sooner or later corrections were made and accepted everywhere, though the north Pacific coastlines of America and Asia were not accurately and completely defined until after 1800. Mapping of the American interior started more slowly, but the main features of the two continents had all been incorporated into maps by about the same time.

As areas of the unknown faded from world maps, it became impossible to doubt that modern men had surpassed the ancients in matters of geography. The same thing happened to the heavens when, in 1609, Galileo used his newly invented telescope to see the mountains on the moon and to discover the moons of Jupiter. During the ensuing eighty years, astronomical discoveries came thick and fast and quite eclipsed anything new arriving in Europe from America and other distant parts of earth. A brilliant and convincing synthesis of new astronomical data was achieved in 1687, when Isaac Newton published his famous book, *Principia Mathematica*. This book, in effect, bridged the gap between the heavens and the earth by submitting both to the same laws of motion, and provided Europeans with a universal, mathematical, and predictive model of what science ought to be.

The Americas had nothing whatever to do with this remarkable development of European thought, but it is interesting to realize that when in the nineteenth century the Newtonian world machine encountered its first serious intellectual challenge, America returned to the limelight, playing a critical, central role in the emergence of the evolutionary world view that prevails today. It all began with botany, which underwent a remarkable expansion in the sixteenth and seventeenth centuries, before astronomy usurped primacy in European consciousness. From Columbus's time onwards, ships could easily carry seeds and cuttings back to Europe, where a learned establishment already existed whose members were eager to learn about new and curious plants. Early in the seventeenth century, some of them conceived the idea of recreating the Garden of Eden by assembling all the different plants of earth in a great botanical garden. These enthusiasts tried hard to make new species of plants grow on European soil and in many instances, they succeeded. As the variety of plants known to the learned botanists of Europe multiplied in this fashion, they faced the question of how to classify them all. The inherited scheme, deriving from Theophrastus' (c. 372–287 B.C.) writings about plants, proved to be entirely inadequate. Eventually a new system was introduced by Carl Linnaeus in 1753 which is still in use. Like the cartographers' newly-perfected world maps, it became an essential guide to the world's flora; and proved again that moderns had left the ancients behind when it came to understanding the world.

Other forms of life were difficult to transport by sailing ship and far harder to acclimate for study in Europe. Initial reports by inexpert observers propagated much misinformation and provoked much confusion. Only when learned Europeans decided it was necessary for them to visit strange and distant parts in order to see things for themselves did reliably sifted and relatively complete information about animals and birds, together with geological studies of the New World become available. The voyage of Alexander von Humbolt, 1799–1803, followed by twenty years devoted to writing and publishing the new data he had gathered, constituted the first great landmark of this advance. Even more significant was the voyage of Charles Darwin on the Beagle, 1831–37. Darwin needed a little more than twenty years for digesting his observations before he published his famous book *The Origin of Species* in 1859, thus radically challenging older notions about how God had created the natural world. Taken together, the work of Humbolt and Darwin constitute a remarkably clear example of how careful study of American data, and prolonged effort to fit the novelties into existing knowledge, induced a basic transformation of inherited understandings and a rather reluctant recognition of the universality of change.

Similar adjustment became necessary in other realms of knowledge too, though less dramatically so, and with stronger carryover from pre-Columbian idea systems. The classification of humankind derived from the Bible meant that natives of the New World had to descend somehow from Shem, Ham, or Japheth. This was patently implausible, but if American Indians did not fit the scheme, the Bible itself would be discredited. Euro-

peans were extremely loathe to admit such a possibility, and debate about the descent of American Indians was not seriously joined before Darwin published his book in 1859. What Europeans did instead was to prefer a classification of humankind derived from pagan Greeks, whereby peoples were classified as savage, barbarous, or civilized. This left plenty of room for discussion as to where American Indian peoples fitted into the pattern, and was, in fact, the way in which Europeans made sense of the human scene in the New World until the 1850s. Thereafter, Darwinian ideas about human evolution opened up debate on the Biblical story of human origins, making the question of how American Indians and all the other new-found peoples of the earth could possibly be traced back to Noah particularly critical. Simultaneously, Lewis Henry Morgan took a closer look at the customs of the Iroquois, thereby helping to originate social anthropology and inaugurating a far more sympathetic understanding of the cultural differences that separated whites from surviving Indian peoples.

Other aspects of human affairs and the ideas that directed them tended to run strongly one way, from Europe towards the new colonial domains. Thus Spanish viceroys and lesser officials strove hard to fit New World realities into the legal forms prescribed by the Catholic church and by the royal government of Spain. Other colonists did similarly, though with less rigorous central direction. Discrepancies could not be avoided, nor completely disguised, as we saw in the case of peonage and slavery. But in general, the Americas were swiftly and successfully absorbed into an expanding European system of law and international practice. This was especially true of transoceanic trade and of the war and diplomacy that enveloped it. But the same held true of the law enforced in local courts within colonial jurisdictions. Together with Christianity, propagated far and wide among the Indians by pioneering missionaries, European authorities drove most of the high culture of Indian peoples underground and destroyed many Indian institutions entirely. Carryovers from the pre-Columbian past were possible at a family and village level, and remain very much alive in parts of rural Mexico, and in Central and Andean America as well. But for the most part the Indians were subjected and their culture fragmented and distorted by the outside pressure of imported European institutions and ideas.

Sure of their own superiority, white immigrants and their descendants adhered to the aspects of the European heritage that worked reasonably well in the new environment, and modified what did not, sometimes by borrowing something from the Indians – like moccasins or maize – but more often by improvising and inventing ways of cutting corners.

Nevertheless, for Europe and the expanding western world as a whole, the fact that a familiar legal and social system prevailed with only marginal modifications, does not invalidate Elliott's claim that "in discovering America, Europe discovered itself." For the contrast between themselves and the native inhabitants of America was a constant reminder of the fact that European civilization was only one among a number of different ways of life. To be sure, easy successes in the New World reinforced the ethnocentrism that had previously reigned undisputed everywhere. But just because they were sure their own ways were the best, Europeans could afford to be curious and examine other people's behavior with a modicum of detachment and respect. At first this capability had more scope in Asia and especially in European encounters with Japan and China than it did in the Americas. But the ethnic encounters of the New World also contributed to the eventual, more sensitive exploration of cultural pluralism, beginning with the work of Lewis Henry Morgan.

America's impact on Europe's consciousness of itself was far stronger in a different sense, for, as we just saw, the exploration and exploitation of America swiftly demonstrated beyond any reasonable doubt that contemporaries knew more and could do things that their predecessors had been incapable of. By the mid-eighteenth century, the idea of progress – in knowledge and technol-ogy if not in virtue – became difficult to deny. This still remains fundamental to our consciousness. We expect advances in technology and new discoveries in science. We no longer think that history records an irreversible falling away from Eden or some classical golden age. We expect truths to emerge from experience, not from ancient texts. We are, in short, the heirs of all that has happened to transform human outlooks and expectations since 1492 when the shape of the world and patterns of human interaction underwent the sharpest turn of recorded history.

In this moment of quincentenary commemoration, it is worth pondering how much has flowed from the opening of the oceans to regular contacts. Assuredly the world will never be the same again, and, as the story of Columbus and the consequences of his voyages so vividly show, the way in which human hopes and purposes tangle with the evolving ecosystem and with the system of meanings that routinely govern collective behavior will continue to produce unforeseen and surprising results. History is always surprising. We understand it most imperfectly. Reflecting on how unexpectedly Columbus's achievement transformed human life both in the Old World and the New is therefore a good way to prepare for the further surprises that undoubtedly lie ahead.

What triumphs and disasters will the next five hundred years produce? We will have to wait and see.

WILLIAM H. MCNEILL
University of Chicago, Emeritus

ENCOUNTERING THE NEW WORLD
1493–1800

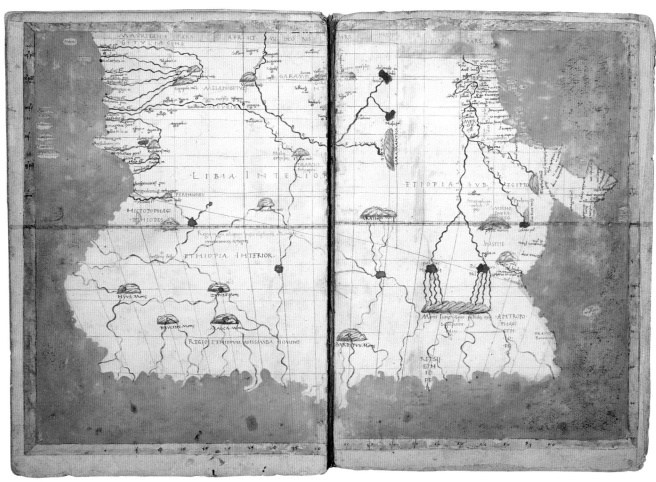

FIG. I

[Africa IV.] In: [Claudius
Ptolemy. "ATLAS OF FIFTEEN
MAPS." [Italy? ca. 1450.]

  The owner of this atlas,
which was originally drawn
in about 1450, has updated
the map of Africa. With a
swash of blue paint the news
of Portuguese discoveries is
recorded, obliterating the
Ptolemaic conception of the
original delineation.
(Item no. 6).

ENCOUNTERING THE NEW WORLD
1493–1800

T HE EUROPEAN DISCOVERY of the Americas was a geographical event of unparalleled magnitude. The publication in 1493 of Columbus's letter to the Spanish court announcing the discovery of what he believed to be a new route to the Eastern spiceries, was the beginning in Europe of the realization that the perimeters of the "known" world were changing dramatically and irrevocably.

The encounter of the Old World with the New was the first event of world-altering importance for which the appropriate technology was in place for relatively rapid dissemination of information about its many facets. Some old theories and legends appeared to come true with the discovery; others were abandoned in favor of realities that were even more astounding than the fruits of the medieval imagination. It became apparent that pictures were needed to emphasize the novelty of the American experience and to make it understandable, and the publishers and printmakers of Europe quickly began to fill the void.

The prints, maps, and books in this exhibition make it evident that people of the sixteenth and seventeenth centuries had the same love of adventure, fear of the unknown, and curiosity about the world as people today. *Encountering the New World, 1493–1800*, reveals literally how Europeans pictured the Americas.

I. *Beginnings*

THE MEDIEVAL ORDER

The world of medieval Europe was circumscribed by boundaries of many kinds. A strict social hierarchy preserved order among men; walled cities protected those within from hostile neighbors and marauding strangers; and according to the most authoritative views, the habitable world of Europe, Asia, and Africa was surrounded by an ocean that separated civilization from the threat of the unknown. The monsters of the seas and the beastmen of legend occupied the borderlands—the very edges

of man's mental map of his world. Beyond the boundaries lay chaos; within, a precarious order ruled.

Mediterranean sailors developed both the nautical chart (portolano) and written sailing directions (routiers or rutters) as practical guides to known waters. To them, the Atlantic was the edge of an outer ocean that marked the western boundary of the habitable world as they knew it. But to seafaring men of northwest Europe the Atlantic had long been a highway used by fishermen, by Celtic missionaries in search of peoples to proselytize, by hermits in need of solitude, and by the land-hungry seeking new territory to colonize. By the beginning of the seventh century A.D., Scotland, the Orkneys, and the Shetlands had been settled; before A.D. 800, Iceland; and by A.D. 1000 Norsemen had set eyes on Labrador.

Stories of islands to the north circulated throughout Europe in unwritten sagas and in manuscript, but since the northern peoples were not in close cultural contact with southern centers of learning, this information was not incorporated into the body of accepted geographical "fact." However, the existence of islands in the Atlantic was hardly unsettling, for medieval mapmakers had always placed them in the fringe of ocean that lay along the western edge of the habitable world.

In the fifteenth century, the Portuguese began to explore systematically the west coast of Africa and to settle Atlantic island groups such as Madeira and the Azores. When Bartolomeu Dias rounded the southern tip of Africa in 1487 and entered the Indian Ocean, theoreticians were forced to consider the failings of a Ptolemaic geography that had presented that body of water as a land-locked sea. (Fig. 1).

Had the new geographical information been merely an advance in theory, it would probably have remained the province of scholars and cosmographers, who debated such things. But this new knowledge had practical implications of the highest order, for the water route around Africa gave Portugal the means to enter the lucrative spice trade with the East Indies. No longer was it necessary for her to be subject to the demands of the endless stream of middlemen who traditionally controlled the trade on the long overland route through Asia to distribution centers in the Italian city-states.

A WIDER WORLD

Although Portuguese success in utilizing exploration to develop new trading contacts with the East was obvious and enviable, Spain was in no position to undertake costly exploratory ventures until the country was unified as a Christian nation-state under Ferdinand and Isabella, the crowns of Aragon and Castile. After the defeat of the Moors at Granada in 1492, Spain was able to look outward, and Christopher Columbus, who had long tried to interest a backer in his plan to sail west to reach the East, appealed to the Spanish sovereigns at an opportune time.

Columbus made landfall on an island in the Bahamas that the native inhabitants called Guanahani (the actual location of which is the subject of lively debate even today), where he was met by people "who went about naked without shame." (Fig. 2). To the questions of where he was and who these people were the Discoverer did not hesitate in answering that he was on an island off the shore of Asia, and that the people were, undoubtedly, Indians from the Spice Islands (even though they were not quite as he had expected them to be). That neither answer was correct became apparent to others, if not to Columbus himself, fairly soon. Columbus's confusion is indicative of the difficulty Europeans initially faced in assimilating new information and relating it to their conventional world-view.

Christopher Columbus would much rather have encountered sophisticated orientals than the "children of nature" who met him with guilelessness, generosity, and a few gold ornaments. But his missionary enthusiasm and his practical outlook made him see the positive side, observing, "how easy it would be to convert these people ... to make them work for us." (Fig. 3).

However, the native population was not always docile. When Columbus returned on his second voyage he discovered that the countrymen he had left behind had apparently been killed by those same Indians who had greeted them with such timidity on his arrival. In time, the Spanish began to differentiate between two tribes that inhabited the area, the gentle Arawaks and the aggressive Caribs. It has been suggested that this distinction was not necessarily founded upon perceived ethnographic differences, but was instead based on the Amerindians reaction to the Spaniards—passive Indians were Arawaks, and hostile Indians were Caribs.

Of the ideological equipment that Europeans brought with them to America, religion was the most important. More than color or national sentiment, it was religion that made Europeans feel distinct from and superior to those peoples they met outside their own continent. Centuries of Christian crusades to free the Holy Land from the infidel had solidified this attitude and placed the question of religion uppermost in the minds of Europeans who came in contact with alien cultures. The Indians of America presented a special problem. Were they innocents, ripe for conversion, living in a Garden of Eden as before the Fall, or were they castaways of God ruled by the Devil? Matters of great practical importance concerning the treatment of the Indians hinged on the answers to these questions. (Fig. 4).

From long tradition Europeans accepted the existence of strange animals and bizarre semi-human creatures. (Figs. 5, 6). As the boundaries of man's "mental map" of his world were extended in this era of discovery, many of those monstrosities migrated west. Explorers told "true tales" of encounters with unusual peoples and animals, and while many of these stories were consciously embellished or invented, it is a fact that communication between newcomers and natives was rudimentary at best, giving rise to numerous possibilities for misunderstanding on both sides. Then, too, observations were often filtered through the screen of traditional European imagery, and many people saw what they expected to see. (Figs. 7, 8).

THE NEW WORLD EMERGES

Incorporating the lands of the western hemisphere into the medieval view of the world posed serious problems for the cartographer. Columbus claimed he had found an off-shore Asian island while others, like Amerigo Vespucci, proclaimed it a New World. (Fig. 9). Fragmentary and conflicting information filtered back to the centers of Europe, and mapmakers faced the task of fitting the pieces together. (Fig. 10). Beginning as the dream of Asia realized, North and South America gradually came to be viewed by some as a massive barrier to the riches of the East. (Fig. 11). Magellan's discovery of a southern passage to the Pacific was an event of the greatest significance, but the difficulty of this route convinced other nations that the location of a passage in more northern latitudes was of the utmost importance. (Figs. 12, 13). By the middle of the sixteenth century this had become a prime target of New World exploration.

FIG. 2

[Columbus's landfall in the New World.] In: Christopher Columbus. DE INSULIS INVENTIS. [Basel, 1493.]

This illustrated printed announcement of Columbus's voyage, popularly called a "Columbus Letter," shows the Discoverer landing on Guanahani in an oared galley.

It is thought that the woodcut may have seen previous service (without the Amerindians) as an illustration in a book depicting Mediterranean ports, for printing and illustration were not inexpensive processes and craftsmen often "made-do" with what was on hand. Nevertheless, the picture presents the "idea" of landfall; the individuality of this event is expressed by the blockcutter's addition of an unclothed native population. (Item no. 8).

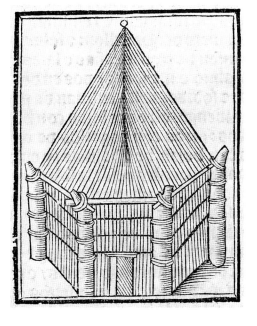

FIG. 3

[Indian dwellings.] In: Gonzalo Fernández de Oviedo y Valdés. CORONICA DELAS INDIAS. Salamanca, 1547.

These woodcuts of Indian huts are among the earliest pictures of buildings on the island of Hispaniola (present-day Haiti and the Dominican Republic) where Columbus established La Navidad, the first European settlement in the New World. Oviedo (who brought the first sugar plants from Spain to America) arrived in the West Indies in 1513 and served the New World Spanish empire in various administrative posts for thirty-four years. (Item no. 9).

FIG. 4
Pices, serinam, reliquam annonam ustulandi ratio. In: Theodor De Bry. GRANDS VOYAGES, PART II (LATIN). Frankfurt, 1609.

It is inaccurate to generalize about "American Indians" as though the Western Hemisphere contained only a single unified culture. Nevertheless, many early writers propagated stereotypes due to their ignorance of the diversity of New World peoples. If one tribe ate strange foods by European standards, the nature of all Indians was brought into question.

Certain elements in the Amerindian diet raised the suspicion that they were in association with devils, a natural consequence of the belief that one acquired the characteristics of the food one ate. This seemed to be indicated by the fact that many native peoples consumed snakes, lizards, toads, grubs, insects, and other "filth." With horrified fascination it was noted that Amerindians "esteem serpents as we do capons," and that serpents were eaten instead of bread.
(Item no. 12).

FIG. 5

[Cannibals.] In: Lorenz Fries. USLEGUNG DER MER CARTHEN. Strasbourg, 1525.

Cannibals, called anthropophagi in Classical texts, were traditionally located at the edge of the habitable world. Although some real cannibalism did exist in certain areas of the Americas (for purposes and with meanings that are very difficult to interpret), the reality often merged with the old legends.

The text below the illustration reads: The cannibals are hideous, horrible creatures, appear to be equipped with dogheads, ghastly to look at, and they dwell on an island discovered by Christophel Dauber ["Little Christopher Dove", i.e., Christopher Columbus] in recent years. (Item no. 13).

FIG. 6

[Dragon.] In: [Samuel de Champlain.] "BRIEF DISCOURS." [France, ca. 1599–1600.]

The author of this manuscript (attributed to Champlain) was a careful observer, but he sometimes became entangled in the preconceptions of a European bestiary that included all manner of imaginary creatures.

The figure shown here he described thus: "There are also dragons of strange figure, having the head approaching to that of an eagle, the wings like those of a bat, the body like a lizard, and only two rather large feet; the tail somewhat scaly, and it is as large as a sheep; they are not dangerous, and do no harm to anybody, though to see them, you would say the contrary." (Item 42b)

FIG. 8

Psittacus Paradisi ex Cuba. In: Mark Catesby. THE NATURAL HISTORY OF CAROLINA, FLORIDA AND THE BAHAMA ISLANDS. Vol. I. London, 1771.

Explorers were especially impressed by the parrots of the West Indies and South America, which were larger and more brightly colored than the African species that were already known in the courts of Europe. Columbus mentioned these vividly colored birds in his "Letter" of 1493, and Brazil was characterized on many early maps as "the Land of Parrots." Native featherwork was brought back to Europe, admired as much for the beauty of the plumage as for the intricacy of the finely worked designs. (Item no. 20a).

FIG. 7
Gottfrid Bernhard Goëtz.
AMERICA. Augsburg
[ca. 1750].

Here, the allegory of
America joins European
representations of the three
continents of the Old World.
AMERICA contains many of
the elements one finds
expressed in the written

accounts of early explorers.
The arrival of Columbus
in the New World in search
of wealth and converts is
pictured on the left. "America"
appears in the foreground as a
dark-skinned, regal woman.
Parrot in hand, enthroned
upon an alligator, she offers
the riches of the country to the
new arrivals. The strange,
exotic background evokes the

sense of dream and wonder
that permeated the accounts
of early travelers.
(Item no. 17).

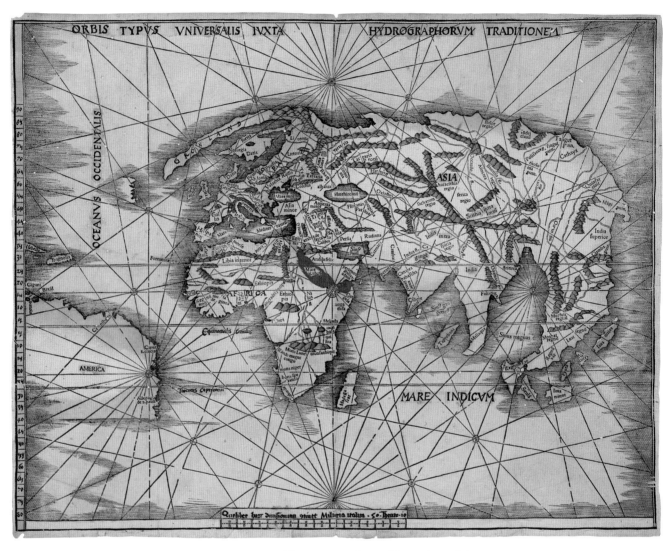

FIG. 9
[Martin Waldseemüller.]
ORBIS TYPUS UNIVERSALIS.
[Nuremberg? ca. 1507–
1513].

The word "America" first
appeared in 1507 in a
pamphlet by the German
cosmographer Martin
Waldseemüller in which he
suggested that the New World
be named for Amerigo
Vespucci—the other

continents were named for
women and it was only fitting
that this one be named for the
man who discovered it.
Waldseemüller's confusion
about the priority of the
discovery was due to
Columbus having claimed
only that he had found a new
route to the East Indies,

whereas Vespucci stated that
he had set eyes on a new
world.

This map, prepared about
1507, is one of the very earliest
to use the word "America."
It has been suggested that
Waldseemüller later realized
his mistake, for the name is
no longer present on the map
he drew in 1513. (Item no. 21).

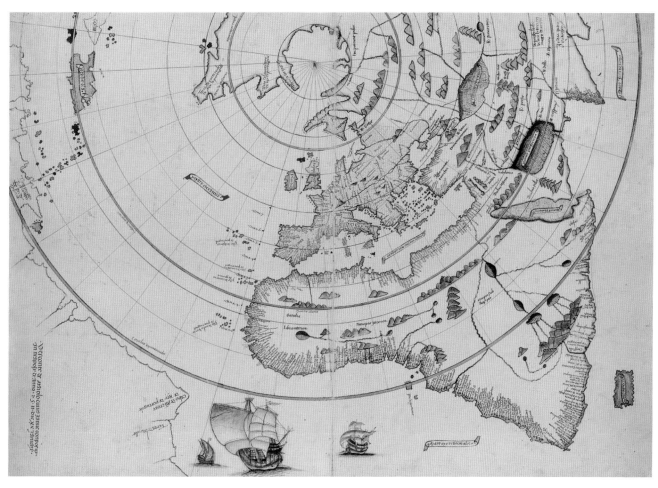

FIG. 10

[Map of the world.] In:
Vesconte Maggiolo.
"PORTOLAN ATLAS."
Naples, 1511.

This world map is an
important early depiction of
European explorations in
America. Maggiolo's place
names include "Lands found
by Columbus," "Land of the
English," and "Land of the
Corte Real and of the King of
Portugal." Although the
cartographer confronts certain geographical problems—
North America is identified
with eastern Asia—he avoids
others, such as the exact
configuration of the connection between the northern
and southern continents of
America. (Item no. 22).

FIG. 11
[Western Hemisphere.] In:
Jan ze Stobnicy.
INTRODUCTION IN
PTHOLOMEI
COSMOGRAPHIAM.
Cracow, 1512.

This hemisphere, copied
from the insets on a large wall
map drawn by Martin
Waldseemüller in 1507,
attempts to cope with peren-
nial cartographic problems—
the incorporation of new
information into a standard
body of accepted "truth,"
and the consolidation of
small-area sketches and
verbal information into a
broad picture of the world.

While the mapmaker
presents a Columbian view
of the discovery (America is
an off-shore Asian landmass),
he shows North and South
America as having a con-
tinuous coastline—an
inspired guess, since at that
time no one had yet sailed
along the entire coast.
(Item no. 23).

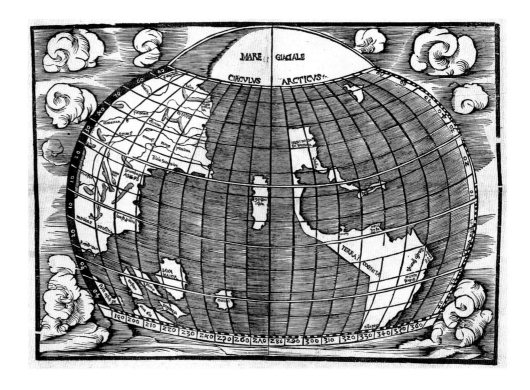

FIG. 12
[Giacomo Gastaldi.] TOTIUS
ORBIS DESCRIPTIO. [Venice,
between 1562 and 1570.]

FIG. 13
Giacomo Gastaldi.
UNIVERSALE
DESCRITTIONE... PAULO
FORLANI VERONESE FECIT.
[Venice, ca. 1565].

For over two decades
Giacomo Gastaldi was the
leading mapmaker of Italy,
both for his technical expertise in execution and for the
information he assembled
and interpreted. The maps
shown here (both of which
were in circulation during the
same period of time) demon-
strate the problems faced by
the cartographer who had to
reconcile often vague or con-
tradictory information from
many sources. One shows a
narrow body of water, the
"Streto di Anian," between
Asia and America and one
does not. The strait now
named for Vitus Bering was
not "discovered" until 1728,
but Gastaldi had guessed its
existence nearly two centuries
earlier. (Item nos. 27 and 28).

10 ENCOUNTERING THE NEW WORLD

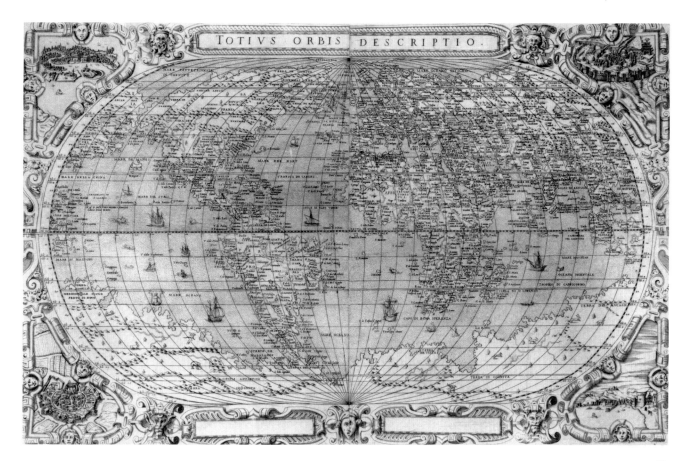

12.

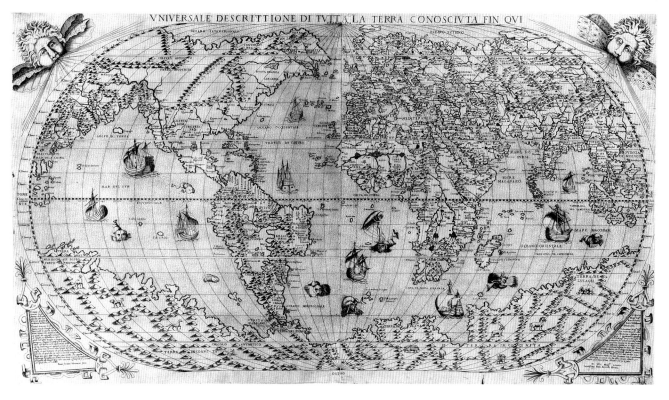

13.

II. *Expanding Horizons*

THE WONDERS OF MEXICO

When Hernán Cortés and his band of conquistadores set foot on the Mexican mainland in 1519 it became obvious to them fairly quickly that this was not a land of simple people in a New World Eden. (Fig. 14). Even to European eyes, the Aztecs were "civilized," and all the more confusing because of it. (Figs. 15, 16). The initial puzzlement of the Spaniards, however, did not long deter them from their overriding goals—gold and glory. (Fig. 17). Cortés's vision of a new Spanish Empire, in combination with the Church's mission of conversion, led inevitably to the shattering of indigenous Mexican culture. (Figs. 18–20).

SOUTH AMERICA REVEALED

In order to settle rival claims and to prevent future conflict between Spain and Portugal, Pope Alexander VI (the Spanish Pope) issued the Papal Bull of 1493 that established Spain's sovereignty over all lands west of a longitudinal line drawn 100 leagues west of the Azores. With seeming prescience, Portugal did not accept this arrangement, although seven more years were to elapse before Pedro Cabral "discovered" Brazil. Spain consented to a new agreement, made without papal intervention, that relocated the demarcation line 370 leagues west of the Cape Verde Islands, leaving to Portugal whatever territory might be east of the line. This was done in the treaty concluded at Tordesillas, Spain, in 1494, which effectively divided South America between Spain and Portugal. (Figs. 21, 22).

LOWLAND RAINFORESTS AND ANDEAN RICHES

Obsessed by the widely-accepted theory that gold and other precious metals and gems were to be found in southern latitudes, European explorers confronted the rivers and steaming rainforests of the Amazon seeking legendary golden cities and fabulous wealth. (Figs. 23–28). In the early 1530s, Francisco Pizarro succeeded in conquering the Inca empire in Peru. Throughout the sixteenth century, the combined wealth of Mexico and Peru poured into the coffers of Spain to make her the richest and most powerful country in Europe. (Figs. 29–31).

FIG. 14
[Tenochtitlan] In: NEWE ZEITTUNG VON DEM LANDE. DAS DIE SPONIER FONDER. [Augsburg? 1522?]

It was reported that when the Spanish soldiers first saw the Aztec capital of Tenochtitlan they said "it was like the enchantments they tell of in the legend of Amadis," and that some asked "whether the things that we saw were not a dream."

This woodcut, the first published view of that city (before its destruction by Cortés), looks like a medieval European town. It is an attempt on the part of the German illustrator to interpret written accounts, the only information available to him, of this "advanced civilization." The faucet-like extensions from the city walls were an attempt to show the causeways that linked this "city in a lake" with the surrounding land. (Item no. 34).

FIG. 15
[Aztec religious ceremonies.]
In: Diego Valadés.
RHETORICA CHRISTIANA.
Perugia, 1579.

An Aztec temple is shown here as the center of activity for Mexican Indian civilization. Dancing, burial customs, cooking, astronomical observation, fishing, worship, and even human sacrifice are all represented in this composite view.
(Item no. 35).

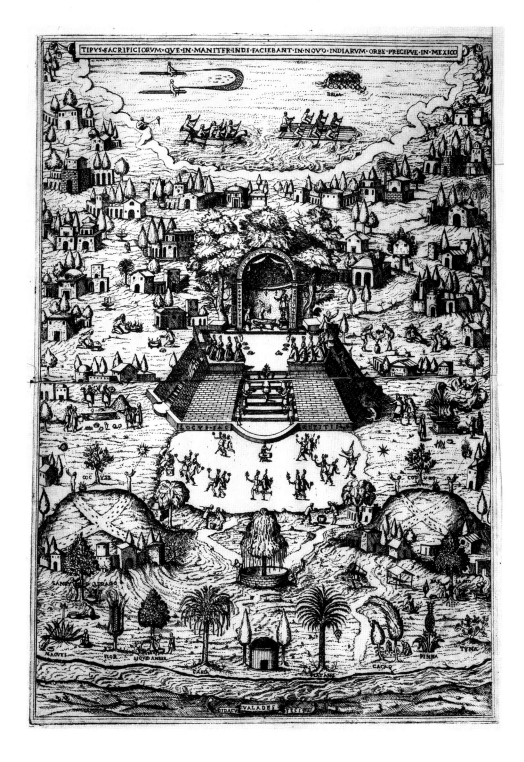

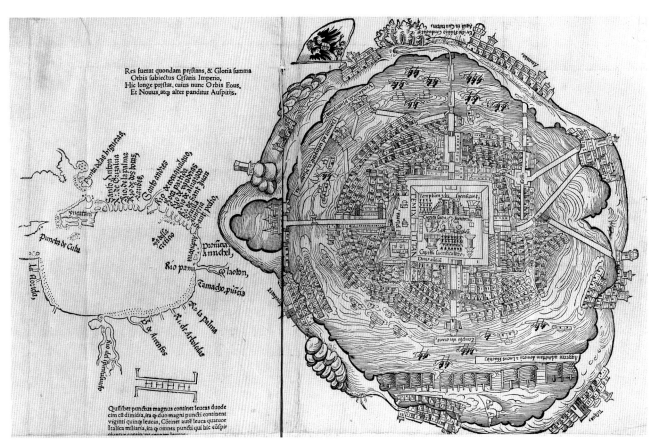

FIG. 16
*[Plan of Mexico City & the
Gulf of Mexico.]* In: Hernán
Cortés. PRAECLARA . . . DE
NOVA MARIS OCEANI.
Nuremberg, 1524.

This plan of Tenochtitlan
as the conquistadors saw
it, attributed to the artist
Albrecht Dürer, is the earliest
known map of an American
city. It was sent to Spain and
published in the first Latin
edition of Cortés's second
"Relation," or report, of
1524 in which he described
his New World activities to
the Spanish crown.

Tenochtitlan was com-
pletely destroyed by the
Spanish in 1521 and Mexico
City, designed by Cortés, was
built upon its ruins. This plan
shows the city before its
destruction: the principal
temples of the Aztecs occupied the main square, causeways connected the city with
the mainland, and an aque-
duct supplied fresh water.
(Item no. 36).

FIG. 17

Labor Vincit Omnia. In: José de Rivera Bernárdez. DESCRIPCION BREVE DE LA MUY NOBLE, Y LEAL CIUDAD DE ZACATECAS. Mexico, 1732.

The illustration shown here is rich with the symbolism of Spanish conquest that would have been understandable to any reader of the day.

Four soldiers, two representing the Old World and two representing New World conquest, support the globe under the dominion of Philip II of Spain ("PLS II"). The sun at the left and the moon at the right signify that the sun never sets on the Spanish Empire. The bundles of arrows represent the kingdoms of Aragon, Navarre, Granada, and Castile, unified at the end of the fifteenth century. The shell represents St. Iago of Compostela (St. James), the patron saint of Spain during the reconquest of the Iberian peninsula. Tales of Spanish victories over incredibly superior forces of Indians led people to believe that St. James had also interceded on behalf of the Europeans in Mexico. The Virgin Mary protects and oversees the enterprise—with determined effort the entire world will become Christian under one prince. Labor Conquers All. (Item no. 37).

FIG. 18

[*Huitzilopochtli, the hummingbird, principal god of the Aztecs.*] In: Juan de Tovar. "HISTORIA DE LA BENIDA DE LOS YNDIOS." Mexico, 1585. (Item no. 41a).

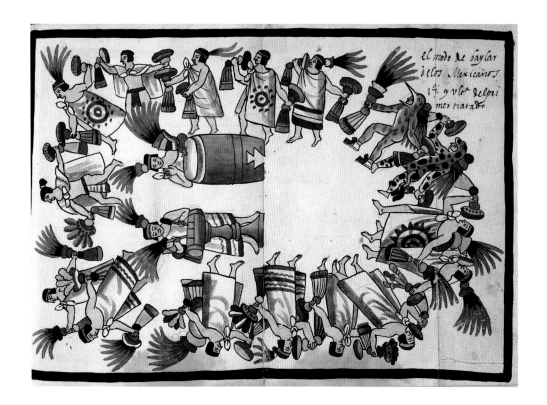

FIG. 19
[*Ritual dance performed by Aztec nobles and priests.*] In: Juan de Tovar. "HISTORIA DE LA BENIDA DE LOS YNDIOS." Mexico, 1585. (Item 41b).

FIG. 20
[*An eagle devouring a bird, the sign by which the wandering Aztecs knew where to establish their city of Tenochtitlan.*] In: Juan de Tovar. "HISTORIA DE LA BENIDA DE LOS YNDIOS." Mexico, 1585.

In an attempt to extinguish idolatry, many Spanish missionaries made a concerted effort to remove all traces of native religious practices, which they accomplished all too well. With hindsight, however, the Church realized that in destroying the records of Aztec civilization they had also deprived themselves of a deeper understanding of that culture. Without such information the clerics could not probe behind Indian "conversions of convenience" to uncover lingering remnants of paganism.

By the middle of the sixteenth century, some missionaries were actively encouraging Spanish-schooled Indians to record what they remembered of Aztec civilization. Juan de Tovar's manuscript history is accompanied by watercolors that illustrate the native religion as it existed before the conquest. The paintings were possibly done by an Indian artist who was drawing on direct recollection of destroyed works.
(Item no. 41c).

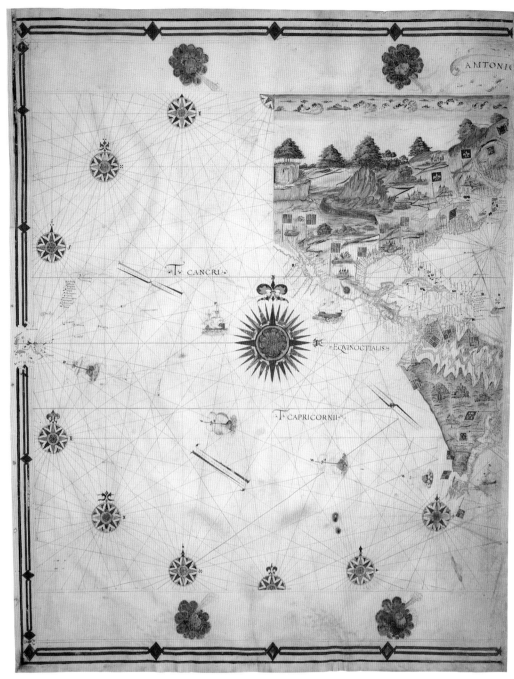

FIG. 21
António [Pereira.] "[NORTH
AND SOUTH AMERICA.]"
[Portugal, ca. 1545].

This is one of the earliest
maps to show the results of
Spanish exploration of the
interior of North and South
America. On the continent of
South America is recorded
Francisco de Orellana's great
trek between 1539 and 1542.

Starting at Quito, Orellana
traveled across the Andes
to the headwaters of the
Amazon and down that great
river to its mouth.

Originally in three parts,
this map has been attributed
to António Pereira, a Portuguese seaman. The other two
sections have not, as yet, been
found. (Item no. 49).

FIG. 22
Ala espada y el compas.
Mas. y mas. y mas. y mas.
In: Bernardo de Vargas
Machuca. MILICIA Y
DESCRIPCION DE LAS
INDIAS. Madrid, 1599.

With his left hand resting
on his sword and his right on
a compass that divides the
world, this conquistador
exemplifies the idea that there
is no limit to Spanish possibilities in the New World.
"With the sword and the
compass, more and more and
more and more" is its message.

The book it introduces is a
practical, "how-to" manual
for the would-be conquerer
written by a soldier whith
many years of experience in
America. (Item no. 50).

FIG. 23
[Landscape.] In: Matthaeus
Merian. GRANDS VOYAGES,
PT. XIV – (GERMAN). Hanau,
1630.

The land and the beasts
that inhabited it took on an
almost malevolent character
as explorers battled their way
through a country that
seemed unfit for "human"
habitation and far away
indeed from the civilized
world. (Item no. 53).

FIG. 24
[Cannibal scenes.] In:
Theodor De Bry. GRANDS
VOYAGES, PT. III –
(GERMAN). Frankfurt, 1593.

Tales of cannibalistic tribes
quickly caught the European
imagination, but even
contemporary observers
criticized the tendency to
show Amerindians in the
midst of hanging human
parts as if they were in a
butcher shop, stating that the
artists had never seen what
they were trying to depict.

Nevertheless, the pictures
circulated widely and served
to emphasize the strangeness
of the world opening up
across the Atlantic Ocean.
(Item no. 54).

FIG. 25

Portrait du Combat Entre les Sauvages. In: Jean de Léry. HISTOIRE D'UN VOYAGE FAICT EN LA TERRE DU BRESIL. Geneva, 1580.

War is one of the most complex of human institutions, with particular characteristics that vary from culture to culture. European observers usually had little comprehension of Amerindian tribal war, which often was highly ritualized and fought for limited ends. (Item no. 55).

FIG. 26

Francois Carypyra. and Louis Marie. In: Claude d'Abbeville. HISTOIRE DE LA MISSION DES PERES CAPUCHINS EN L'ISLE DE MARAGNAN. Paris, 1614.

A "before and after" picture of Christian conversion. Native Americans were taken back to Europe on a regular basis beginning with Columbus, not always voluntarily. Shown here are members of an Indian delegation from the coast of Brazil brought to Paris in 1613, when the French were attempting to establish a colony in what was considered Portuguese territory and sought to forge an alliance with the local population.

Some members of the delegation sickened and died in France. Those who survived were baptized and given Christian names. (Item no. 57).

FIG. 27
[Alligator and Snake]
In: Maria Sybilla Merian.
OVER DE VOORTTEELING EN
WONDERBAERLYKE
VERANDERINGIN DER
SURINAEMSCHE INSECTEN.
Amsterdam, 1719.

In 1699 Maria Merian and her daughter travelled unaccompanied to Surinam, where she concentrated on painting the indigenous plant and insect life. Merian's accuracy, technical expertise, and vibrant artistic skill, along with her noteworthy research in natural history, earned her a solid reputation in scientific circles. (Item no. 59a).

FIG. 28
[Pineapple.] In: Maria Sybilla
Merian. DISSERTATIO DE
GENERATIONE ET
METAMORPHOSIBUS
INSECTORUM
SURINAMENSIUM.
Amsterdam, 1719.

Columbus first encountered
the pineapple on his second
voyage to America in 1493.
Its reception in Europe was
enthusiastic: "The fruit of the
pine-apple . . . undoubtedly
surpasses all the fruits with
which the world is at present
acquainted." (Item no. 59b).

FIG. 29
Americaner in Peru. In:
Hermann Fabronius. DER
NEWEN SUMMARISCHEN
WELT-HISTORIE.
Schmalkalden, 1612.

Depicted here is a Peruvian
Indian. The poem in German
below the image states that
although the people of the
country are naked, its rivers
flow with gold, which Spain
carries away in the holds of
her ships. (Item no. 63).

FIG. 31
*Indi Hispanis aurum
sitientibus.* In: Theodor De
Bry. GRANDS VOYAGES, PT.
IV. (LATIN). Frankfurt, 1594.

In retribution for Spanish
greed, Amerindians are
shown obliging the invaders'
craving for gold by pouring
the molten metal down their
throats. (Item no. 66).

FIG. 30
[Titlepage.] In: Theodor De Bry. GRANDS VOYAGES, PT. VI. (LATIN). Frankfurt, 1596. Theodor De Bry's view of Indians working in the mines. (Item no. 65).

AMERICAE
PARS SEXTA.
SIVE
HISTORIÆ AB HIERONYMO BEZONO
Mediolanêse scriptæ, sectio tertia, res no
minus nobiles & admiratione plenas con
tinens, quàm præcedentes duæ. In hac
enim reperies, qua ratione Hispani opule-
tissimas illas Peruâni regni provincias oc
cuparint, capto Rege Atabaliba: deïde orta
inter ipsos Hispanos in eo regno civilia bella.
Additur est brevis de Fortunatis insulis Cômenta
riolus in duo capita distinctus.
Item ad ditiones ad singula Capita Histo-
riam illustrantes.
Accessit Pervâni regni chórographica Tabula.
AD
INVICTIS RVDOLPH:H:ROM:IM:AVG:
Omnia elegantibus figuris in æs incisis expressa
à Theodoro de Bry Leod:cive autem Fracofurtesi
A° MD XCVI.
Cum privilegio s. c. Ma^tis.

III. *The Lure of North America*

IMPRESSIONS OF NATURAL BOUNTY

The experience of Spain in Mexico and Peru seemed to prove the currently-held European theory that mineral wealth was located primarily in southern latitudes. Those lands were pre-empted by Spain and Portugal, however, and the countries of northern Europe with imperial ambitions were excluded from any possibility of sharing in the bounty by rightful possession. Circumstances forced them to look northward.

No golden cities and "advanced" civilizations awaited explorers who set foot on the Atlantic coast of North America. But, after initial disappointment, the French, the Dutch, and the English, who were active in this region, began to appreciate the possibilities of a country that was extraordinarily rich in natural resources and also free from the extreme climate of the southern lands. The northern Amerindians seemed to live a simple, unstructured life, nurtured with ease by a fruitful land in a climate similar to that of Europe. That these initial perceptions were simplistic and often incorrect would be the source of hardship and of future conflict between the newcomers and the native populations. (Figs. 32, 33).

RESOURCES, DEVELOPMENT, AND TRADE

It became apparent that North America had something to offer other than precious metals. Resources such as wood and fish were in high demand in Europe, and trade with the Amerindians could be developed for furs. Outposts established to oversee these businesses could serve also as plantations to grow crops desirable to the Old World economy. (Figs. 34, 35). Coastal settlements were envisioned as extensions of European empire and as way-stations for expeditions in the ongoing search for a passage through the continent to the wealth of the Indies.

Early explorers took note of native plants and animals and the ways in which they were used by the Amerindians, especially for medicinal purposes. (Figs. 36, 37). The perceived similarity in climate between North America and northern Europe allowed for the possibility of cross-cultivation—plants unknown in Europe could be brought from the Western Hemisphere for cultivation in the Old World, and crops in short supply in Europe could be grown extensively in America.

The European governments that organized or sponsored New World settlement were interested primarily in the economic value of colonies. (Figs. 38–41). Although there were important exceptions, the profitable exchange of goods and the exploitation of natural resources were the primary motives that sent men across the oceans and made their contact with the land and its inhabitants continuous and systematic.

Economic development of the New World required settlers, and all of the European colonial powers promoted immigration. (Fig. 42). The vast tracts of undeveloped land created opportunities at various places and times for whole communities of the persecuted or the dissatisfied to emigrate. For some—English Catholics in Maryland; Spanish and Portuguese Jews in Dutch Brazil and in New York; and Protestant sectarians in many places—America was, above all, a refuge. (Fig. 43).

FIG. 32
Comme les Dames de Dasamonguepeuc Portent Leurs Enfans. In: Theodor De Bry. GRANDS VOYAGES, PT. I. (FRENCH). Frankfurt, 1590.

FIG. 33
Sumendi Cibum Modus. In: Theodor De Bry. GRANDS VOYAGES, PT. I. (LATIN). Frankfurt, 1590.

Some ordinary Amerindian customs struck the newcomers as strange enough to require special comment. The "odd" way that women carried their children is described in detail, as is the "barbarous" custom of sitting on the ground to eat. In this illustration from DeBry (Fig. 33), the engraver has altered the position of the legs shown in John White's original drawing to one more "comfortable" by European standards.
(Item nos. 71 and 72).

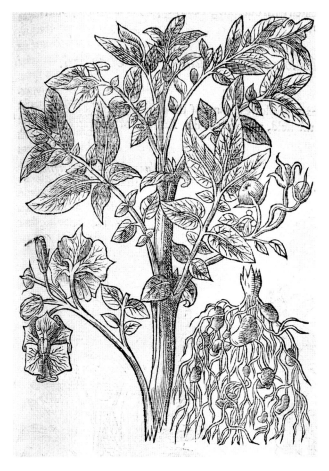

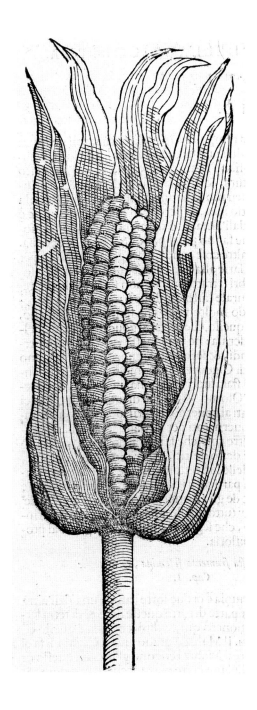

FIG. 34

Of Potatoes of Virginia. In: John Gerard. THE HERBALL OR GENERALL HISTORIE OF PLANTES. London, 1597.

This is the first published picture of the white potato (as distinguished from the sweet potato, also of American origin), a plant that eventually became a staple in European diets.

In 1585 one variety of potato was brought to England by Sir Francis Drake, who had encountered the tubers in his raids on Spanish outposts in Panama. (Item no. 78).

FIG. 35

[Maize.] In: Giovanni Battista Ramusio. TERZO VOLUME DELLE NAVIGATIONI ET VIAGGI. Venice, 1606.

"If maize were the only gift the American Indian ever presented to the world," it has been said, "he would deserve undying gratitude," for it has become one of the most important foods for men and livestock. By the end of the seventeenth century, it was widely grown as a staple in Spain, France, Italy, and in Africa. However maize was introduced to most Europeans by the Turks after its spread east across the African coast of the Mediterranean and was commonly called "Turkey" or "Turkish" wheat. (Item no. 46).

FIG. 36
[Buffalo.] In: Francisco López de Gómara. PRIMERA Y SEGUNDA PARTE DELA HISTORIA GENERAL DE LAS INDIAS. Zaragoza, 1553.

No animal of North America made a more profound impression on Spanish explorers than the bison. Totally unknown to Europeans, they were found in huge herds on the great plains between the Mississippi and the Rio Grande. Descriptions by the chroniclers of some of the early Spanish expeditions are vivid: "and they have between the two Horns an ugly Bush of Hair, which falls upon their Eyes, and makes them look horrid."
(Item no. 47).

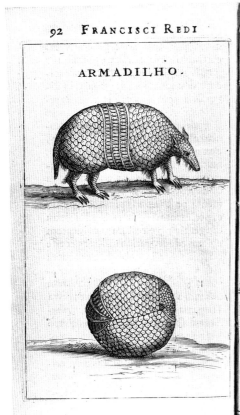

92   FRANCISCI REDI

ARMADILHO.

Pag. 93.

Pece Muger, five piscis ἀνθρωπόμορφος.

FIG. 37
[Armadilho.] and *[Pece Muger.]* In: Francesco Redi. EXPERIMENTA CIRCA RES DIVERSAS NATURALES. Amsterdam, 1675.

Medicinal uses were found for both real and imaginary animals. The bones of the armadillo's tail were ground into a powder and used as a painkiller; some claimed that deafness could be cured as well.

This mermaid ("pece muger," or "pesci donne") from the coast of Brazil reportedly had "tender bones, which [when used in the proper manner] promote chastity, inhibit carnal desires, render men impotent, and are useful in many other ways for the good of the body."
(Item no. 48).

    This illustration, by the
English poet and artist
William Blake, shows
"White" Europe supported
by the resources of "Black"
Africa and "Indian" America.
Vast amounts of labor were
required to ensure that the
wealth of the New World
flowed regularly and in ever-
increasing amounts back to
the homeland.
(Item no. 85).

Europe supported by Africa & America.

London, Published Dec.r 1.st 1792, by J. Johnson, St Pauls Church Yard.
80

FIG. 39
Indianer können der Spanier.
In: Theodor De Bry. GRANDS
VOYAGES, PT. IV. (GERMAN).
Frankfurt, 1594.

Early in their occupation
the Spaniards had to face the
fact that Indians were
difficult to enslave. Among
other problems, it was too
easy for them to escape back
into their surrounding
societies. Moreover, the
goodwill of Indian tribes was
often a necessity for the

Europeans' survival. The
Spanish looked to Africa to
solve the labor shortage, and
the Atlantic slave trade
began, in which the Dutch,
French, Portuguese, and
English were also active
participants.

In the tradition of the
"Black Legend" of Spanish
cruelty in the New World, this
illustration shows Amer-
indians committing suicide
rather than face enslavement
by the Spanish. (Item no. 86).

FIG. 40
[Heretics.] In: [Samuel de Champlain.] "BRIEF DISCOURS." France, ca. 1599–1600.

Although this manuscript account of a voyage to the Spanish west Indies in 1599, attributed to Samuel de Champlain, focuses mainly upon the indigenous flora and fauna, the author gives some information about the lot of the Indians under Spanish rule. According to him, the harshness of the Inquisition caused the Indians to flee to the hills. The Spaniards who followed them were captured and eaten. At the time of this writing the situation had eased somewhat, and the author states that had the Spaniards continued still to chastise the Indians according to the rigor of the Inquisition, they would all have died by fire. (Item no. 42a).

FIG. 41

[Processing sugar cane.] In: Johann Ludwig Gottfried. NEWE WELT UND AMERICANISCHE HISTORIEN. Frankfurt, 1655.

From its fifteenth-century beginnings with the establishment of Portuguese trading posts on the west coast of Africa, the Atlantic trade in African slaves spread quickly throughout the American colonies. Slavery itself was an age-old, world-wide institution, and the particular forms it took in the colonization of America were due to a variety of factors. One thing is certain, the demand for labor to produce sugar, tobacco, silver, and other commodities was great, and slaves brought from Africa supplied most of it.

Reduced to a commodity, slaves were truly "faceless" people, depicted as part of the furniture of colonial life. They typically figured in illustrations of the time in much the same way that tools might appear in a farmyard scene. (Item no. 87).

This figure representeth the Engine, to wind off the silk from the Cods, with Furnaces and Cawdeerns necessary thereto

FIG. 42
This figure representeth the
engine, to wind off the silk
from the cods. In: [Edward
Williams.] VIRGINIA'S
DISCOVERY OF SILKE-
WORMES. London, 1650.

Denied easy access to Asia,
the English attempted to
establish a silk industry in
Virginia. The colonists
greatly underestimated the
difficulty of the process,
however, and at first mis-
understood even the kind of
mulberry leaf the silk worms
fed upon. Despite high hopes,
hardly any silk was ever produced in colonial Virginia.
(Item no. 92).

A. La ville Royale de Melilot.   B. La grande Eglise.   C. Le Palais du Parakousse ou Roytelet.   D. La montagne d'Olaimy.
E. Le Temple du Soleil.   F. La Figure de la Plante Sensitive et de la Fleur.

FIG. 43

Paysage de la Province de Bemarin au Royaume d'Apalache. In: Charles de Rochefort. HISTOIRE NATURELLE ET MORALE DES ILES ANTILLES DE L'AMERIQUE. Rotterdam, 1681.

The author of this book was a Huguenot who actively encouraged French Protestants to settle in America.

Shown here is the apocryphal city of Melilot, capitol of the Indian kingdom of Apalache. Although usually assigned a location in present-day Florida, it was sometimes placed in Georgia. According to Rochefort, the kingdom was home to a mysterious band of European Protestants who had been driven from their Virginia colony. This is probably the first printed illustration of a North American utopian city.
(Item no. 99).

FIG. 44

Wie die Wilden in Florida. In: Theodor De Bry. GRANDS VOYAGES, PT. II — (GERMAN). Frankfurt, 1591.

A meeting between Frenchmen and Florida Indians in 1564 is depicted here with a European court atmosphere.

The French brought ready-made marker columns with them to be used in establishing their claims to New World territory. The Indian "king," Athore, shows the French commander, René de Laudonnière, a column left by the leader of an earlier reconnaissance mission in 1562. Misunderstanding its significance, the Indians have treated it as an idol and have brought tribute to honor the place. The gestures of the worshippers are European, however, rather than Amerindian. (Item no. 100).

FIG. 45

[Cannibal America.] In: Ferdinando Gorges. AMERICA PAINTED TO THE LIFE. London, 1659.

The poem engraved on the verso of this print makes it obvious that the cannibal image symbolizes not only the Amerindians, but also the malevolent aspects of the land itself.
(Item no. 104).

IV. *Settlement Brings Conflict*

CONFLICTS OF CULTURE

Most Europeans came to the New World convinced of their religious and cultural superiority and were often ill-equipped to understand or to accept the cultures they encountered. Early explorers attempted to follow traditional Old World diplomatic practices in their contact with Amerindian societies, especially those that were perceived to be monarchical. This often consisted of identifying a "king" and persuading him to swear allegiance to a more powerful Christian, European monarch. Many times hierarchies were assumed that did not, in fact, exist within Indian societies. (Fig. 44).

Throughout the period of discovery and exploration, illustrations of Amerindians were often used in symbolic fashion to underscore the perceived differences between Old and New World cultures. Subtly or blatantly, these pictures mirrored European fears and biases. Indians were variously depicted as threatening—cannibals who lured men to their doom, forces of anarchy that undermined the foundations of society—or, sometimes, as children who craved the guiding hand of European civilization. (Fig. 45).

Compared with the English, the French appeared to be more successful in their dealings with the North American Indians. The traders and "coureurs du bois" that ranged through the interior were not immediately followed by large numbers of colonists in search of land to settle, reducing opportunities for direct conflict with Amerindian culture (in the period of early contact, at least).

The missionaries sent out from France to win souls earned the respect of the native population by their willingness to live alongside them and to share in their lives. The Catholic religion they brought with them possessed a rich iconography and liturgy that provided a visual focal point for the potential converts, which was, perhaps, more appealing to some Indians than the Bible-based, textual proselytizing of English Protestants.

If French explorers and traders had little immediate impact on the land and its native peoples, the same could not be said for the English settlers that arrived in ever greater numbers throughout the seventeenth century. (Fig. 46). Eventually, all coastal Indian societies found themselves threatened as occupants of land the newcomers found especially desirable, with the result that these tribes were either decimated or pushed to the interior. (Fig. 47).

On numerous occasions Amerindians tried to rid themselves of the unwanted presence of Europeans by attacking their settlements. In the short-term these measures were often successful, insofar as they slowed the advance of settlement for a time. King Philip, chief of the Wampanoags, a tribe in southern New England, led an uprising against the English colonists in 1675. After initial successes the Indians gradually succumbed to the settlers, and in August, 1676, King Philip was himself trapped and killed. (Fig. 48).

CONFLICTS OF EMPIRE: CHALLENGES TO SPAIN

Colonial rivalries were especially intense in the Caribbean. Although war between Spain and England was not declared until the attack of the Spanish Armada in 1588, there had been hostilities from the time of Queen Elizabeth's accession to the throne thirty years before; the attacks of English sea rovers (most notably Sir Francis Drake) on Spanish colonies and shipping in America were a constant source of ill will between the two powers. (Figs. 49–50). Dutch depredations were largely the result of the war between Spain and the Netherlands (1572–1648), the New World focus of which was Dutch trading and settlement rights in the Caribbean.

Sixteenth-century religious wars, in the wake of the Reformation, spread across the Atlantic. The earliest French settlements in North America were sponsored by Huguenots (French Protestants) between 1562 and 1568 near what is now Jacksonville, Florida. Catholic imperial Spain believed that this settlement of heretics was installed dangerously close to the route the plate fleet followed on the first leg of its cross-Atlantic voyage and took immediate steps to destroy the young French colony. (Fig. 51).

In those tumultuous times, both state sponsored and unofficially encouraged privateering against "enemy" shipping often degenerated into unlicensed piracy for private ends.

VYING FOR NORTH AMERICA

Almost as soon as the Hudson River was explored in 1609, the Dutch arrived to trade in furs, establishing posts at Fort Orange (Albany), Fort Nassau (near Philadelphia), and New Amsterdam. However, these trading settlements were not supported by extensive immigration, an important factor in their eventual overthrow by the English in 1664. (Fig. 52).

Unlike the English, whose settlements hugged the coast, French traders and missionaries immediately pushed toward the interior of the country in search of furs and converts. (Fig. 53). While the rivers were highways for daily business, there was also the hope that there would eventually be found a water route leading through the continent to the Pacific Ocean.

Rivalry between France and England escalated throughout the eighteenth century, but not all skirmishes between the two powers took place on the battlefield. (Fig. 54). Cartography provided a public forum for questions of empire that was less bloody, perhaps, but no less intense than military action. (Figs. 55, 56).

INDIAN ALLIES

A great stir was caused in English court circles in the spring of 1710 when four "Indian Kings" from North America were presented to Queen Anne. The political aim was evident: seeing the splendors of London and the power of England might help ensure the continued support of the Five Nations (the Mohawks, Oneidas, Onondagas, Cayugas, and the Senecas, later joined by the Tuscaroras to constitute the Six Nations), who had sustained heavy losses over the years because of their faithful allegiance to England. (Fig. 57). John Verelst was commissioned to execute the official portraits, and the symbolism used in his paintings creates a gallery of virtues for these allies of the English — fortitude, courage, craftiness, and statecraft. The paintings were copied by engravers, and the pictures circulated widely. (Fig. 58).

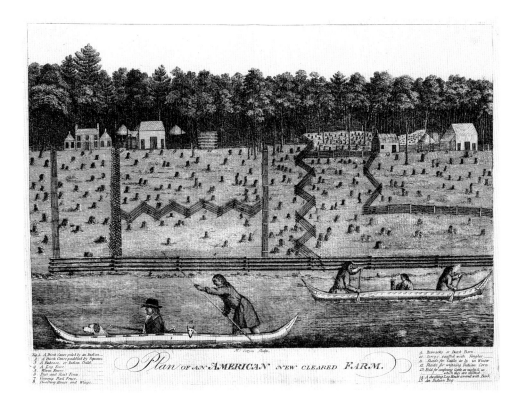

Fig.1. A Birch Canoe poled by an Indian.
2. A Birch Canoe paddled by Squaws.
3. A Baboon, or Indian Guild.
4. A Log Fence.
5. Worm Fence.
6. Post and Rail Fence.
7. Virginia Rail Fence.
8. Dwelling House and Wings.
9. Barracks, or Dutch Barn.
10. Bergs' stuffed with Singles.
11. Shade for Cattle to ly in Winter.
12. Shade for winning Indian Corn.
13. Fold for confining Cattle at night, &c. in which they are milked.
14. A dwelling Log House covered with Bark.
15. An Indian Dog.

Plan OF AN AMERICAN NEW CLEARED FARM.

FIG. 46
Plan of an American New Cleared Farm. In: Patrick Campbell. TRAVELS IN THE INTERIOR INHABITED PART OF NORTH AMERICA. Edinburgh, 1793.

The native American population in the north allowed game to roam freely, practiced companion planting of maize and beans, and did not use the same planting area year after year. From the first, however, English settlers cleared the land and fenced their property in accordance with European custom. These practices permanently (and almost immediately) changed the landscape that had so impressed the colonists upon their arrival and were directly responsible for many confrontations between the newcomers and the Indians. (Item no. 111).

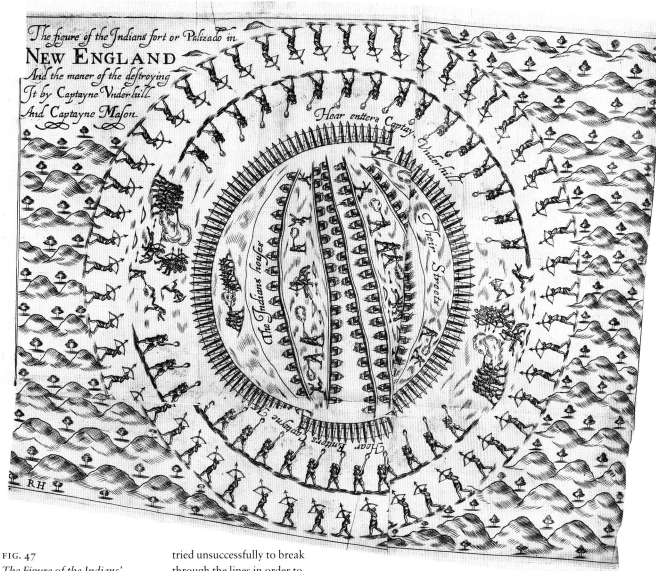

FIG. 47
*The Figure of the Indians'
Fort or Palizado in New
England.* In: John Underhill.
NEWES FROM AMERICA.
London, 1638.

In the opinion of political
leaders in Massachusetts and
Connecticut, the Pequot tribe
was an undesirable neighbor.
This battle plan depicts an
attack that took place in
1637 on a fortified Pequot
village in Mystic, Connecticut, by the English and their
Narragansett Indian allies.
English soldiers with firearms
were arranged around the
inner circle, while the Narragansetts with bows and
arrows occupied the outer
circle. The desperate Pequots

tried unsuccessfully to break
through the lines in order to
avoid certain annihilation.
The Narragansetts were so
horrified by the ferocity of
the English attack, which
included the slaughter of
women and children (less
than five out of 500 escaped),
that they withdrew in shame.
(Item no. 112).

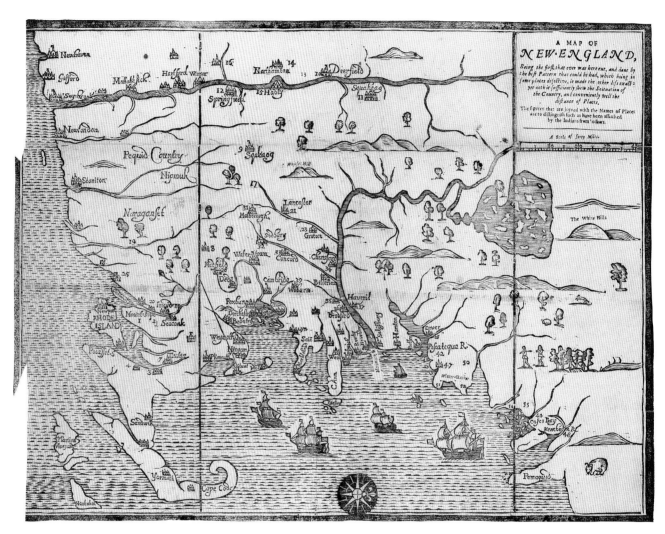

FIG. 48
John Foster. *A Map of New England*. From: William Hubbard. A NARRATIVE OF THE TROUBLES WITH THE INDIANS IN NEW ENGLAND. Boston, 1677.

This map, the first to be drawn, cut, and printed in America, is a "news map" keyed by number to show the battles of King Philip's War. The map is oriented with north at the right. Shown here is the earliest, or "White Hills," version. When it was copied in London later that same year, the woodcutter mistakenly identified the mountains at the right as the "Wine Hills." (Item no. 115).

FIG. 49
[Buccaneer smoking a pipe.]
In: Alexandre Olivier
Exquemelin. HISTOIRE DES
AVANTURIERS. Paris, 1686.

Buccaneers (who were
often the human debris of
earlier expeditions to
America) hunted the wild
cattle that had multiplied
unchecked since the mid-
sixteenth century when large
numbers of Spaniards abandoned their island plantations
to search for treasure in
Mexico and Peru. These men
were superb marksmen,
an advantage when they
joined together to harrass
Spanish shipping.

Here, a pipe-smoking
buccaneer and his dog pause
before his village on their
return from a hunt.
(Item no. 118).

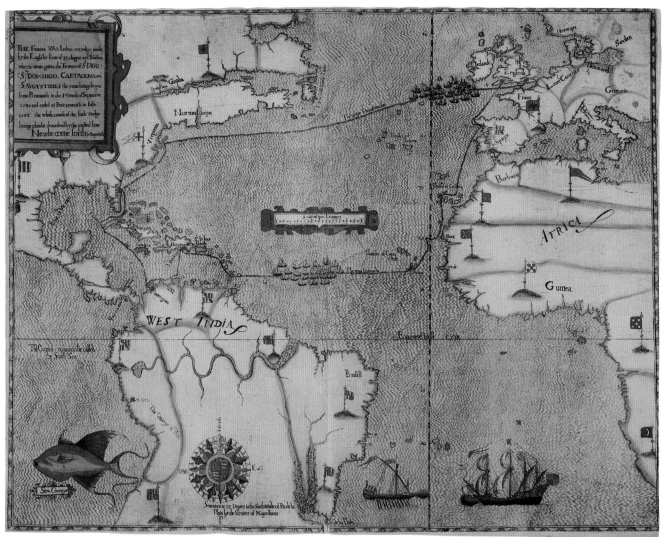

FIG. 50

Baptista Boazio. THE
FAMOUSE WEST INDIAN
VOYADGE MADE BY THE
ENGLISHE FLEET. London,
1589.

Justice-seeking and
recompense for Spanish ill-
treatment of Englishmen were
the stated purposes of Sir
Francis Drake's depredations
on Spain's New World
empire. Success at capturing
Spanish and Portuguese
wealth made him a rich man,
although his expeditions were
often beset with problems—
by accident of timing, cities
yielded no ransom, cornered
prey eluded his grasp, and
treasure slipped through his
fingers. Drake was a great
sailor, however, and the first
Englishman to circumnav-
igate the globe. His daring
elevated him to the status
of hero.

This map traces Drake's
route from the Cape Verde
Islands to Santo Domingo,
Cartagena, and St. Augustine
in an expedition undertaken
against the Spanish in 1585.
On the way home, his fleet
stopped at Virginia to check
on English colonists put
ashore ten months before.
(Item no. 122).

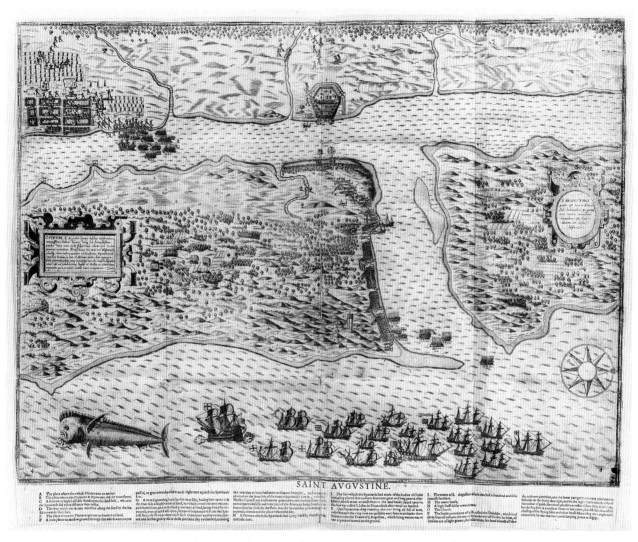

SAINT AVGVSTINE.

FIG. 51

St. Augustine. From: Walter Bigges. A SUMMARIE AND TRUE DISCOURSE OF SIR FRENCES DRAKES WEST INDIAN VOYAGE. London, 1589.

This is the earliest known view of St. Augustine, the first permanent European settlement in what is now the United States. It was established in 1565 by Pedro Menéndez de Avilés as a base of operations to guard Spanish interests in the Caribbean and to eradicate the French Huguenot presence in "Florida." In this last charge Menéndez was successful: the French settlement was destroyed and the colonists were massacred. It is reported that he placed the following inscription on the site: "I do this not to Frenchmen, but to heretics."

In 1586 Sir Francis Drake burned the town and fort of St. Augustine, and this illustration was published to accompany the account of his raids on Spanish America. The fish at the lower left is supposed to be a dolphin. (Item no. 127).

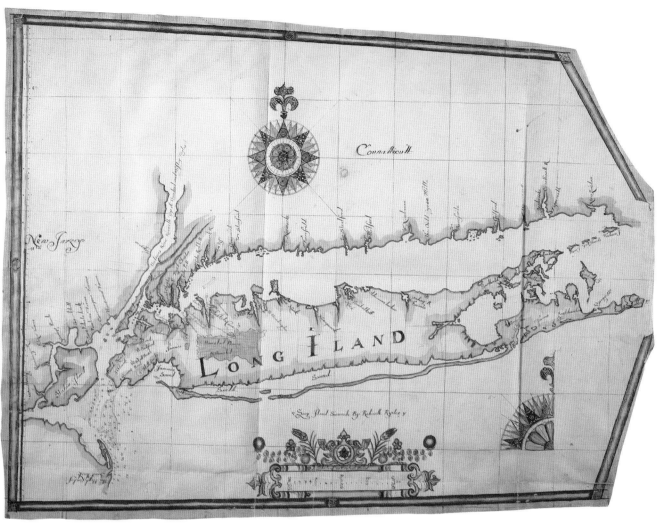

FIG. 52

"Long Island sirvaide by Robartte Ryder." [London, ca. 1679.] From: ["THE BLATHWAYT ATLAS."] London, 1683.

Decorative, generalized maps may have served the needs of armchair travelers, but colonial administrators needed more precise information for the business of government and often contracted with local surveyors to produce accurate reconnaissance maps. This English manuscript map of Long Island, its shape determined by the animal skin upon which it was drawn, is impressive for its beauty as well as for its accuracy. Based upon a draft of the first extensive survey of the area, it was used by English administrators after the defeat of the Dutch in 1673.

(Item no. 130).

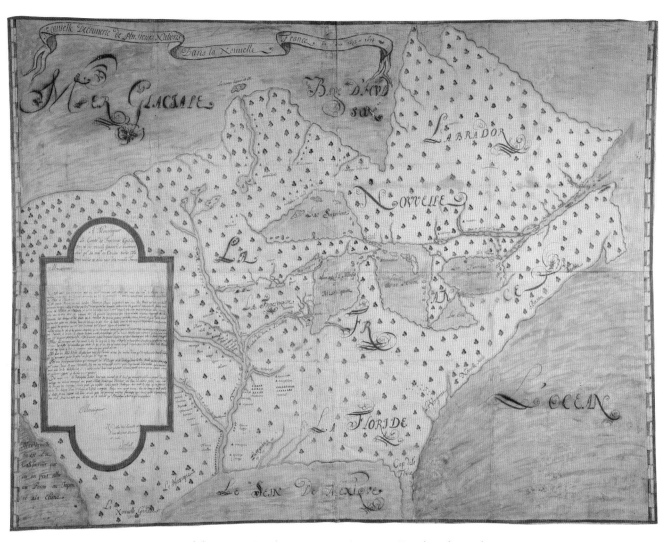

FIG. 53

[Louis Joliet.] "NOUVELLE
DÉCOUVERTE DE PLUSIEURS
NATIONS DANS LA
NOUVELLE FRANCE." [New
France, ca. 1674].

In 1673 the French fur
trader Louis Joliet and Father
Jacques Marquette, a Jesuit
missionary, explored the
Mississippi River, but records
they kept of their trip were
lost in a canoe accident. Upon
his return to Quebec, Joliet
drew a map from memory

and the manuscript shown
here is one of several copies
that were made at the time.
This is one of the earliest
representations of the course
of the Mississippi. Away from
the areas actually traversed
by Joliet, however, it is a
fairly poor geographical view.
Among other things, the map
shows the locations of
various Indian nations,
information that was of vital

interest to French traders and
missionaries.

In the cartouche at the left
is a copy of a letter to the
governor of New France,
Comte Frontenac, in which
Joliet alludes to the fertility
of the country. He also states
that the Missouri River
leads to California and the
Pacific, a convenient (but
nonexistent) water route to
the riches of the Orient.
(Item no. 131).

QUEBEC, *The Capital of* NEW-FRANCE, *a Bishoprick, and*
Seat of the Soverain COURT.

1. The Citadel. 2. the Castle. | 7. Cathedral of Our Lady. | 11. S.t Charles River. | 14. The Bishop's House. 15. The Parish Church of the Lower Town.
3. Magazine. 4. y.e Recolets. | 8. The Palace 9. y.e Seminary | 12. The Common Hospital. | 16. The Upper Town 17. y.e Lower Town.
5. Ursulines. 6. Jesuits. 7. | 10. The Hôtel Dieu. | 13. The Hermitage of the Recolets. | 18. The Platform & Battery of Cannon 19. The Isle of Orleans. 20. Point Lieve.

Engrav'd & Printed By Tho.s Johnston for Step.n Whiting.

FIG. 54

Thomas Johnston. QUEBEC, THE CAPITAL OF NEW FRANCE. [Boston, 1759.]

There are very few views that portray Quebec between its founding in 1608 and the English siege by General Wolfe in 1759. This latter event, however, occasioned many commemorative and "news views." The siege of Quebec was of great interest to English colonists in the north—Bostonians had participated in the military action, for New France was close enough to be perceived as a direct threat.

This print, by the Boston engraver Thomas Johnston, is the earliest American engraved view of Quebec. Although the advertiser proclaimed that it was done "from the latest and most authentic French original," it is actually based on an inset from a French map published forty years earlier.
(Item no. 136).

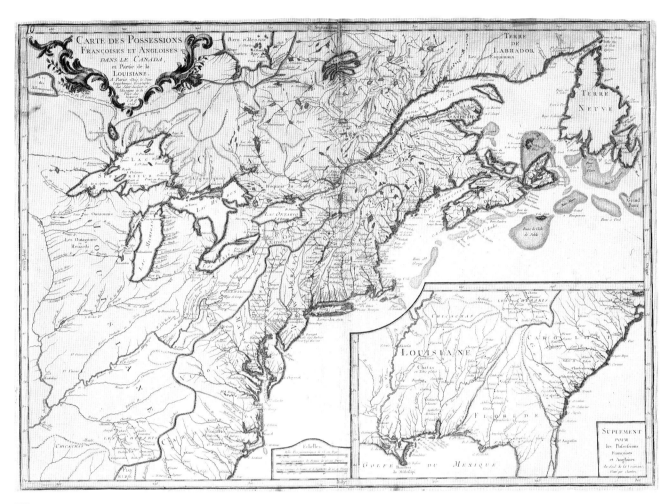

FIG. 55

S. G., Sieur de Longchamp.

CARTE DES POSSESSIONS

FRANÇOISES ET ANGLOISES.

Paris, 1756.

The French view of North
America. (Item no. 137).

FIG. 56
Thomas Kitchin. A NEW AND
ACCURATE MAP OF THE
BRITISH DOMINIONS IN
AMERICA. [London, ca.
1764.]
   The English view of North
America. (Item no. 138).

Sa Ga Yeath Qua Pieth Tow King of the Maquas

FIG. 57
John Verelst. SA GA YEATH
QUA PIETH TOW KING OF
THE MAQUAS. [London,
1710].

Sa Ga Yeath Qua Pieth
Tow, baptized Brant, is
identified as a great hunter.
The scene behind him recalls
his offer to Queen Anne to
run a deer to death. Brant
died shortly after his return
to America, but his grandson,
Joseph Brant, came to
London in 1776.
(Item no. 142).

FIG. 58
THE BRAVE OLD HENDRICK
THE GREAT SACHEM OR
CHIEF OF THE MOHAWK
INDIANS. [London, ca.
1740.]

The Emperor of the Six
Nations, baptized Hendrick,
had a long association with
the English. Hendrick headed
the delegation to London in
1710 and, on his return thirty
years later, was presented
with an elaborate suit
trimmed in lace. He is shown
here as he was painted in
this finery, which was passed
on to his son after his death
(as an ally of the English) at
Lake George in 1775.
(Item no. 147).

The brave old Hendrick the great SACHEM or Chief of the Mohawk Indians,
one of the Six Nations now in Alliance with & Subject to the King of Great Britain.
Sold by Eliz Bakewell opposite Birchin Lane in Cornhill.

v. *Europeans Become Americans*

SETTLING IN

By the middle of the eighteenth century, North America was no longer a land of newcomers who looked upon themselves as Europeans. Many families had been here for two or three generations, and the lands they had settled were no longer outposts of Old World civilization, but bustling towns built by those who now considered North America their home. It is often assumed that American towns "just grew." Many of them, however, were carefully planned, although bursts of economic activity and population growth rapidly obscured the lines and logic of the initial designs. (Figs. 59–62)

BREAKING AWAY

Although British colonists grew to consider North America their home, ties to the Old World were strong. As the colonies developed, their benefit to the mother country increased, and economic and legal disagreements between the mother country and her overseas provinces increased as well. The fire of dissatisfaction that had been smouldering throughout the eighteenth century was fanned by egalitarian theories spawned by the European Enlightenment, and incidents between the colonials and the British government escalated. (Fig. 63) Attempts at conciliation based upon mutual concessions failed, and British troops detailed to destroy weapons stores at Concord became embroiled with the Massachusetts militia at Lexington. Proceeding to Concord the British accomplished their aim, but after a fight at the bridge (the "shot heard round the world") they were forced to retreat, first to Lexington and then to Boston.

The conflict had become a war, and European publishers responded to the requirement for news and information. (Figs. 64, 65). In Germany, Balthasar Leizelt and Francis Habermann added views of American cities to their standard stock of optical views, which were engravings designed to be shown in peep shows (or raree shows) with an apparatus that used candles and mirrors to project the image. (Figs. 66, 67). Printmakers in England, France, and the Netherlands addressed political issues in addition to responding to specific events. Their pictorial assessments of the relationship between England and America give an idea of the ways in which the colonies were perceived by a contemporary European audience.

THE PROMISE OF AMERICA

The Age of Enlightenment was born of a new view of the world, nourished by exploration and tales of travel and supplemented by scientific discoveries and insights into the structure of the natural world. (Figs. 68, 69). In harmony with the ideas of the English philosopher John Locke, the Americans had succeeded by means of a "political contract" in bringing forth a new, comprehensive political structure on republican principles. Having declared their independence on the basis of human rights, the Americans had gone on to fight for independence, and ended by creating a new political community.

Two men, Benjamin Franklin, the pragmatic scientist and "simple" man of the people, and George Washington, a man of dignity, self-control, and innate chivalry, personified the promise of America. In fact, at the close of the eighteenth century, these men were perceived by many to be "America" itself. (Figs. 70, 71). But America was no longer an idea or myth; it was a new country that had to prove itself in terms of actual achievement within the community of states and nations, and many European observers traveled through the new country and reported on its progress and its problems. (Figs. 72–75). If the Age of Enlightenment had made models of the Americans, it was up to the Americans to see that the model became reality. (Fig. 76).

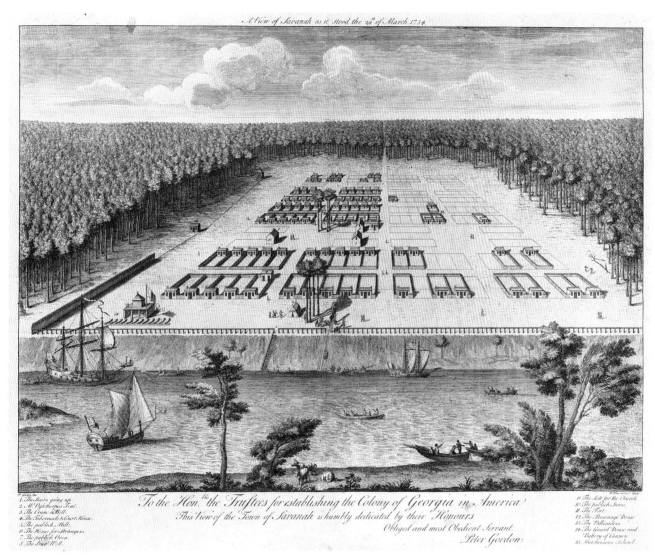

To the Hon.ble the Truftees for establishing the Colony of Georgia in America
This View of the Town of Savanah is humbly dedicated by their Honours
Obliged and most Obedient Servant
Peter Gordon.

1. The Stairs going up.
2. M.r Oglethorpes Tent.
3. The Crane & Bell.
4. The Tabernacle & Court House.
5. The publick Mill.
6. The House for Strangers.
7. The publick Oven.
8. The draw Well.

9. The Seat for the Church.
10. The publick Stores.
11. The Fort.
12. The Parsonage House.
13. The Pallisadoes.
14. The Guard House and Battery of Cannon.
15. Hutchinsons Island.

FIG. 59
Peter Gordon. A VIEW OF
SAVANAH. [London, 1734.]
James Oglethorpe, the
English philanthropist who
founded Georgia, chose the
site and laid out the city of
Savannah in 1733. Intended
as a refuge for debtors,
Georgia was also designed to
be a buffer colony against the
Spanish in Florida.
The original plan called for
a number of wards, each
consisting of forty house lots.
Fronting the square on two
sides were lots set aside for
churches, stores, and other
public uses. Here, Savannah
is shown in its earliest days,
being carved out of the
wilderness. Ten years later
there were about 350 houses,
many of which were considered "large and handsome."
(Item no.150).

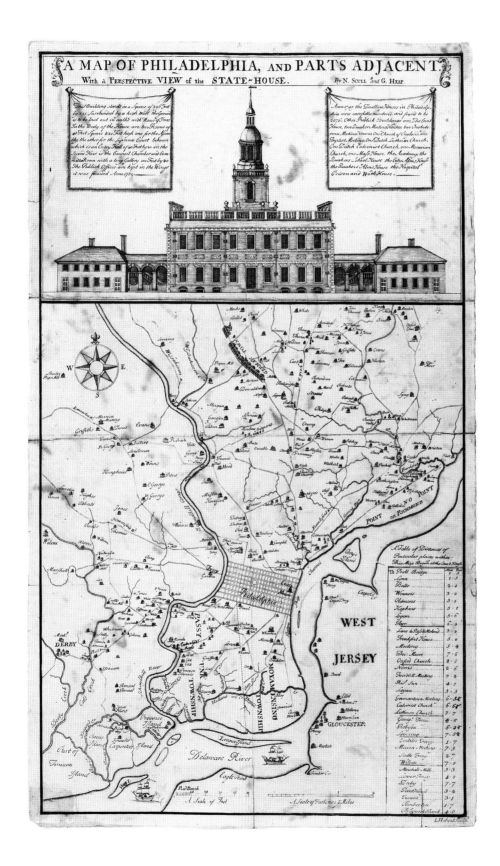

A MAP OF PHILADELPHIA, AND PARTS ADJACENT.
With a PERSPECTIVE VIEW of the STATE-HOUSE. By N. SCULL and G. HEAP.

FIG. 60

Nicholas Scull and George Heap. A MAP OF PHILADELPHIA, AND PARTS ADJACENT. WITH A PERSPECTIVE VIEW OF THE STATE-HOUSE. [Philadelphia, 1752.]

This, the most popular early map of William Penn's "Greene Country Towne," has two claims to distinction. Published in 1752, it is the first printed map to show how Philadelphia developed from the original plan. It also contains the first view of the State House, or Independence Hall. The building had not yet been completed, which accounts for the precarious manner in which Heap drew the as yet unfinished belfry.

By the end of the century, pictures of public buildings often appeared as decorations on maps, gestures of pride in the achievement of those who had made North America their home.
(Item no. 153).

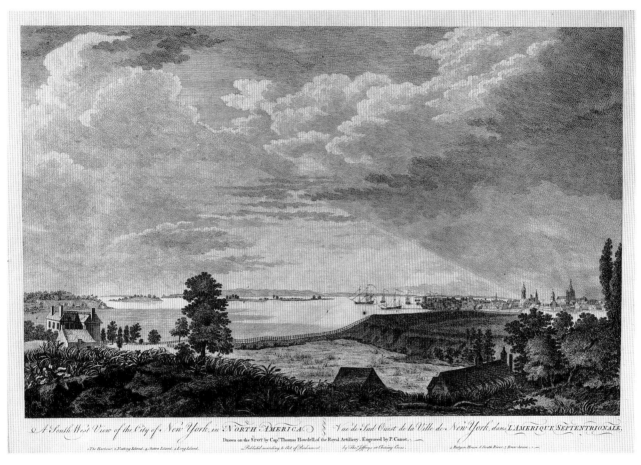

A South West View of the City of New York, in NORTH AMERICA. Vue de Sud Ouest de la Ville de New York, dans L'AMERIQUE SEPTENTRIONALE.

Drawn on the SPOT by Cap.t Thomas Howdell, of the Royal Artillery. Engraved by P. Canot.
Published according to Act of Parliament by Tho.s Jefferys at Charing Cross.
1.The Harbour. 2.Nutting Island. 3.Staten Island. 4.Long Island.   5.Rutgers House. 6.South River. 7.Brew-house.

FIG. 61
Thomas Howdell. A SOUTH
WEST VIEW OF THE CITY
OF NEW YORK. [London,
1763–65].
    This picture of New York
was drawn about 1764 by
Captain Howdell at a site
near the East River presently
occupied by public housing
projects. The particular view
of lower Manhattan chosen
by the artist is now obstructed
by the FDR Drive and the
approaches to the Manhattan
Bridge. (Item no. 156).

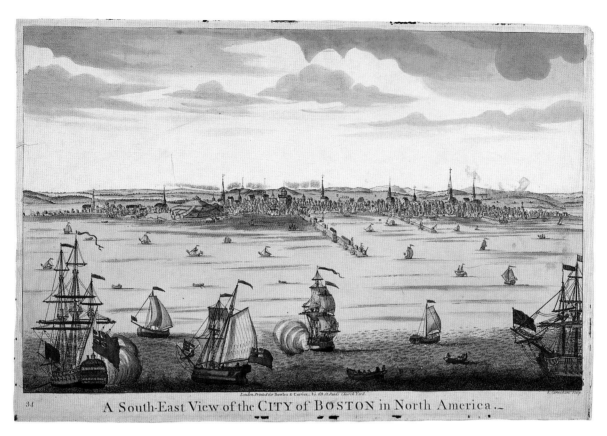

A South-East View of the CITY of BOSTON in North America.

FIG. 62
John Carwitham. A SOUTH-
EAST VIEW OF THE CITY OF
BOSTON IN NORTH
AMERICA. London,
[ca. 1794.]
This print of Boston harbor
was based upon a view by
William Burgis drawn about
1722.
(Item no. 158).

FIG. 63
THE BOSTONIANS PAYING
THE EXCISE-MAN, OR
TARRING & FEATHERING.
London, 1774.
The Liberty Tree provides
a backdrop for this popular
satire depicting the brutal
treatment meted out to John
Malcomb, Commissioner of
Customs, for attempting to
collect duty in Boston.
Malcomb was tarred and
feathered, made to drink
enormous quantities of tea,
and threatened with hanging.
(Item no. 168).

The BOSTONIAN'S Paying the EXCISE-MAN, or TARRING & FEATHERING
Plate I.

REFERENCES

AN EXACT VIEW of THE LATE BATTLE AT CHARLESTOWN June 17th 1775

FIG. 64
Bernard Romans. AN EXACT
VIEW OF THE LATE BATTLE
AT CHARLESTOWN JUNE
17TH, 1775. [Philadelphia,
1775.]

Bernard Romans—civil
engineer, naturalist, cartographer, and historian—was
responsible for an important
group of maps of the
Revolutionary period. In
1774 and 1775 he was in
Boston and engraved this
eyewitness account of the
Battle of Bunker Hill showing
the towns of Boston and
Charlestown. The outcome
of the battle was of interest
to a European audience and
this view was re-engraved in
England for British circulation
in the following year.
(Item no. 171).

FIG. 65

REDDITION DE L'ARMÉE
ANGLOISES. Paris, [ca. 1781.]

On October 19, 1781,
Cornwallis surrendered with
7,000 men. This print was
intended to be used with a
vue d'optique. (Item no. 187).

FIG. 66
Balthasar Frederic Leizelt.
VÜE DE LA NOUVELLE
YORCK. Augsburg, [ca.
1776.]

FIG. 67
Balthasar Frederic Leizelt.
VÜE DE PHILADELPHIE.
Augsburg, [ca. 1776.]

Many optical views of
American towns, such as
these, were adapted from
pictures of European cities.
Nevertheless, their wide
circulation transmitted an
image of urban America to a
vast audience eager for news.
They can best be appreciated,
perhaps, as a contemporary
manifestation of the earlier
tendency to "make do" with
the "idea" of a place rather
than with an accurate
description.

Although these prints claim
to depict New York and
Philadelphia, they are
actually copies of an engraving of Deptford, England—
one half of the original serves
as New York, the other half
as Philadelphia.
(Item nos. 172, 173).

FIG. 68
[George Ehret.] *Magnolia* In:
Mark Catesby. THE NATURAL
HISTORY OF CAROLINA,
FLORIDA, AND THE BAHAMA
ISLANDS. Vol. II. London,
1743.

FIG. 69
Phoenicopterus Bahamensis.
In: Mark Catesby. THE
NATURAL HISTORY OF
CAROLINA, FLORIDA, AND
THE BAHAMA ISLANDS. Vol.
I. London, 1731.
  Mark Catesby, an
eighteenth-century English
naturalist and traveler, strove
to provide a complete and
accurate visual record of the
flora and fauna of the
southeast of North America
and the Bahamas. In a
remarkable tour de force,
Catesby singlehandedly
observed, drew, engraved,
and finally published his
observations in two folio
volumes, with some two
hundred plates. He was the
first to paint North American
birds in their natural settings.
(Item nos. 165a and b).

MOUNT VERNON in VIRGINIA
The Seat of the late Lieut General GEORGE WASHINGTON
Commander in CHIEF of the Armies of the United States

Alexander Robertson Delineavit.

Francis Jukes Sculpsit.

London. Pub.d March 31.st 1800 by F. Jukes N.o 10 Howland Street

and by Al.r Robertson Columbian Academy Liberty Street New York.

FIG. 70
Alexander Robertson.
MOUNT VERNON IN
AMERICA. London, 1800.
    A view of Mount Vernon,
the earliest engraving of
Washington's home to be sold
in Great Britain.
(Item no. 190).

FIG. 71
Amos Doolittle. A DISPLAY
OF THE UNITED STATES OF
AMERICA. New Haven
[1788.]

Six times during Washing-
ton's term as President,
Doolittle issued a version
of this patriotic display,
periodically bringing the
names of states and terri-
tories and their statistics
up to date. This, the first state
of the engraving, shows
Washington in civilian dress;
beginning with the second
state of 1790, all the others
show him in military uniform.
(Item no. 191).

FIG. 72

An American log-house. In: Victor Collot. A JOURNEY IN NORTH AMERICA. Paris, 1826.

General Victor Collot fought under Rochambeau during the Revolution and later became governor of Guadeloupe. In 1796 he returned to North America to survey the Ohio and Mississippi Rivers for the French government. Before his death in 1805 the account of his expedition and an accompanying atlas of maps and plates had been printed, but the unbound sheets were suppressed for political reasons and were not offered for sale until 1826. Shown here is Collot's view of an American log cabin. (Item no. 192).

An American Log-house.

FIG. 73

American stage wagon. In: Isaac Weld, Jr. TRAVELS THROUGH THE STATES OF NORTH AMERICA. 4th edition. London, 1800.

Isaac Weld sailed from Dublin to see if America offered a happier alternative to Europe, and after two years of travel, decided that it did not.

Weld describes a stage-coach: "The coachee is a carriage peculiar, I believe, to America; the body of it is rather longer than a coach but of the same shape. In the front it is left quite open down to the bottom, and the driver sits on a bench under the roof of the carriage. There are two seats in it for the passengers, who sit with their faces towards the horses." (Item no. 194).

American Stage Waggon.

FIG. 74

FIG. 74

[Andrew Ellicot.] PLAN OF
THE CITY OF WASHINGTON
IN THE TERRITORY OF
COLUMBIA. Philadelphia,
1792.

A major cartographic
activity during the last decade
of the eighteenth century was
the mapping of states and
cities. The city of Washington,
D.C., was designed by the
engineer and architect Pierre
Charles L'Enfant, who refused
to release a map for fear of
land speculation. Andrew
Ellicott, an assistant to
L'Enfant, drew the first official plan of the city, which
was published in 1792.
(Item no. 195).

FIG. 75
Second Street North from Market St. wth. Christ Church. In: William Birch. THE CITY OF PHILADELPHIA. Philadelphia, 1800.

The Birch views of Philadelphia provide visual documentation of the beginnings of an American way of life. John Birch executed twenty-eight views of Philadelphia with the intention of recording a complete cross-section of the city that at the time represented the zenith of American culture and achievement. He reviewed both public and private buildings and also included details of street life that had not yet been portrayed in America. A sense of accomplishment and pride is reflected in these images of the American city. (Item no. 198).

SECOND STREET. North from Market S.t wth. CHRIST CHURCH.
PHILADELPHIA.

FIG. 76
AN EMBLEM OF AMERICA. London, 1800.

At the close of the eighteenth century "America" receives a Classical presentation, visual evidence of the anticipated success of the "noble experiment" of American democracy— Reason is in harmony with Nature. (Item no. 200).

An EMBLEM of AMERICA.

ITEM LIST

1.
[Map of the world.]
In: Zaccaria Lilio. DE
ORIGINE & LAUDIBUS
SCIENTIARIUM . . .
CONTRA ANTIPODES.
Florence, 1496.

2.
[Map of the world.]
In Claudius Ptolemy.
COSMOGRAPHIA. Ulm, 1482.

3.
Ulm. In: Hartmann Schedel.
DAS BUCH DER CRONIKEN
UNND GESCHICHTEN.
Augsburg, 1496.

4.
[Map of the World.]
In: Hartmann Schedel. LIBER
CHRONICARUM. Nuremberg,
1493.

5.
Olaus Magnus. [CARTA
MARINA.] Rome, 1572.

6.
[Africa IV.]
In: [Claudius Ptolemy.
"ATLAS OF FIFTEEN MAPS."]
[Italy? ca. 1450.]

7.
Jan van der Straet.
CHRISTOPHORUS
COLUMBUS LIGUR
TERRITORIBUS OCEANI
SUPERATIS ALTERIUS.
[Antwerp, ca. 1575–90.]

8.
*[Columbus's landfall in the
New World.]*
In: Christopher Columbus.
DE INSULIS INVENTIS [Basel,
1493.]

9.
[Indian dwellings.]
In: Gonzalo Fernández de
Oviedo y Valdés. CORONICA
DELAS INDIAS. LA HYSTORIA
GENERAL DE LAS INDIAS.
Salamanca, 1547.

10.
*Indianer da sie wolten
probieren ob die Spanier
unsterbliche.*
In: Theodor De Bry. GRANDS
VOYAGES, PT. IV (GERMAN).
Frankfurt, 1613.

1.

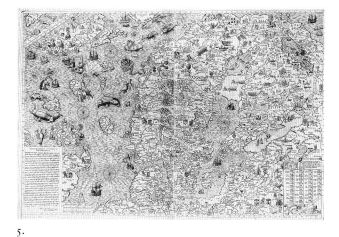

5.

2.

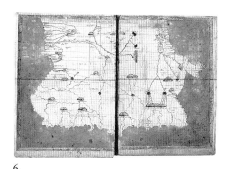

6.

3.

7.

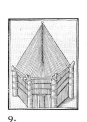

8.

9.

4.

10.

11.
[Eden.]
In: Theodor De Bry. GRANDS
VOYAGES. PT. III, (LATIN).
[Frankfurt], 1592.

12.
*Pices, serinam, reliquam
annonam ustulandi ratio.*
In: Theodor De Bry. GRANDS
VOYAGES, PT. II (LATIN).
Frankfurt, 1609.

13.
[Cannibals.]
In: Lorenz Fries. USLEGUNG
DER MER CARTHEN.
Strasbourg, 1525.

14.
[Blemmyae.]
In: Levinus Hulsius. DIE
FÜNFFTE KURTZE
WUNDERBARE
BESCHREIBUNG. Nuremberg,
1603.

15.
*Quomodo in
Gigantum insula.*
In: Johann Theodor DeBry.
GRANDS VOYAGES, PART X
(LATIN). Oppenheim, 1619.

16.
*Typus Cosmographicus
Universalis.*
In: NOVUS ORBIS
REGIONUM. Basel, 1532.

17.
Gottfrid Bernhard Goëtz.
AMERICA. Augsburg,
[ca. 1750.]

17a.
Gottfrid Bernhard Goëtz.
AFRICA. Augsburg,
[ca. 1750.]

17b.
Gottfrid Bernhard Goëtz.
ASIA. Augsburg, [ca. 1750.]

17c.
Gottfrid Bernhard Goëtz.
EUROPA. Augsburg,
[ca. 1750.]

11.

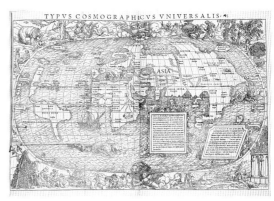

16.

12.

17.

13.

17a.

14.

17b.

15.

17c.

18.
[The Guaiacum.]
In: [Lydia Byam.] A
COLLECTION OF EXOTICS,
FROM THE ISLAND OF
ANTIGUA. [London, 1797.]

Item 19 has been deleted.

Item 20 has been deleted.

20a.
Psittacus Paradisi ex Cuba.
In: Mark Catesby. THE
NATURAL HISTORY OF
CAROLINA, FLORIDA AND
THE BAHAMA ISLANDS. Vol.
I. London, 1771.

Item 20b has been deleted.

21.
[Martin Waldseemüller.]
ORBIS TYPUS UNIVERSALIS.
[Nuremberg ?, ca. 1507–
1513.]

22.
[Map of the world.]
In: Vesconte Maggiolo.
"PORTOLAN ATLAS." Naples,
1511.

23.
[Western Hemisphere.]
In: Jan ze Stobnicy,
INTRODUCTIO IN
PTHOLOMEI
COSMOGRAPHIAM. Cracow,
1512.

24.
Oronce Fine.
COSMOGRAPHIA
UNIVERSALIS AB ORONTIO
OLIM DESCRIPTA. [Venice:
Cimerlinus, 1566.]

25.
Hadji Ahmad. [A COMPLETE
AND PERFECT MAP
DESCRIBING THE WHOLE
WORLD.] [Venice, 1568;
restrike, 1795.]

26.
[MAP OF THE WORLD.]
[Florence? ca. 1552.]

18.

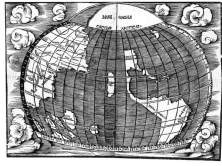

23.

20a.

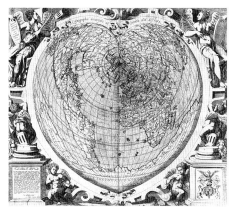

24.

21.

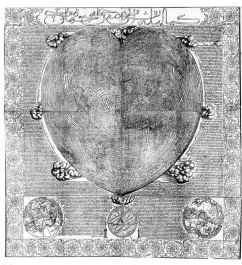

25.

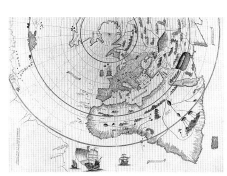

22.

26.

27.
[Giacomo Gastaldi.] TOTIUS
ORBIS DESCRIPTIO. [Venice,
between 1562 and 1570.]

28.
Giacomo Gastaldi.
UNIVERSALE
DESCRITTIONE... PAULO
FORLANI VERONESE FECIT.
[Venice, ca. 1565.]

29.
Antonio Florian. [MAP OF
THE WORLD, IN GORES.]
[Venice, 1566.]

30.
Philippe Buache. CARTE DES
NOUVELLES DECOUVERTES
AU NORD DE LA MER DE
SUD. Paris [1752.]

31.
[Gerhard Friedrich Müller.]
NOUVELLE CARTE DE
DECOUVERTES FAITES PAR
LES VAISSEAUX RUSSES.
St. Petersburg, 1754.

32.
Henry Roberts. CHART OF
THE N.W. COAST OF
AMERICA AND THE N.E.
COAST OF ASIA, EXPLORED
IN THE YEARS 1778 AND
1779. London, 1784.

33.
*Quaratione Indi Sua
Commercia.*
In: Theodor De Bry. GRANDS
VOYAGES, PT. V (LATIN).
Frankfurt, 1595.

34.
[Tenochtitlan.] In: NEWE
ZEITTUNG VON DEM LANDE.
DAS DIE SPONIER FONDER.
[Augsburg? 1522?]

35.
[Aztec religious ceremonies.]
In: Diego Valadés.
RHETORICA CHRISTIANA.
Perugia, 1579.

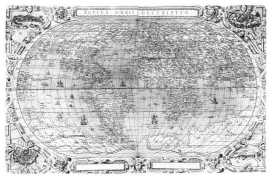

27.

28.

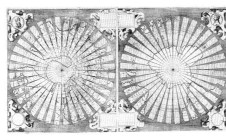

29.

31.

34.

35.

30.

32.

33.

36.
[Plan of Mexico City & the Gulf of Mexico.]
In: Hernán Cortés.
PRAECLARA...DE NOVA MARIS OCEANI.
Nuremberg, 1524.

37.
Labor Vincit Omnia.
In: José de Rivera Bernárdez.
DESCRIPCION BREVE DE LA MUY NOBLE, Y LEAL CIUDAD DE ZACATECAS.
Mexico, 1732.

38.
[Title page.]
In: Theodor De Bry. GRANDS VOYAGES, PT. IV (LATIN).
Frankfurt, 1594.

39.
[Title page.]
In: Theodor De Bry. GRANDS VOYAGES, PT. V (LATIN).
Frankfurt, 1595.

Item 40 has been deleted.

Item 41 has been deleted.

41a.
[Huitzilopochtli, the hummingbird, principal god of the Azrtecs.]
In: Juan de Tovar. "HISTORIA DE LA BENIDA DE LOS YNDIOS." Mexico, 1585.

41b.
[Ritual dance performed by Aztec nobles and priests.]
In: Juan de Tovar. "HISTORIA DE LA BENIDA DE LOS YNDIOS." Mexico, 1585.

41c.
[An eagle devouring a bird, the sign by which the wandering Aztecs knew where to establish their city of Tenochtitlan.]
In: Juan de Tovar. "HISTORIA DE LA BENIDA DE LOS YNDIOS." Mexico, 1585.

42a.
[Heretics.]
In: Samuel de Champlain.
"BRIEF DISCOURS." [France, ca. 1599–1601.]

42b.
[Dragon.]
In: Samuel de Champlain.
"BRIEF DISCOURS." [France, ca. 1599–1601.]

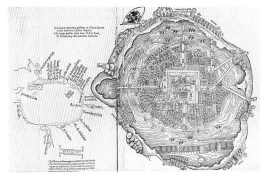

36.

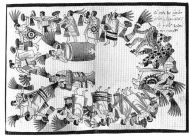

41b.

37.

38.

41c.

39.

42a.

41a.

42b.

42c.
[Rattlesnake.]
In: Samuel de Champlain.
"BRIEF DISCOURS." [France,
ca. 1599–1601.]

Item 42d has been deleted.

42e.
Cochenille.
In: Samuel de Champlain.
"BRIEF DISCOURS." [France,
ca. 1599–1601.]

43.
*The Manner of Propagating,
gathering & curing the Grana
or Cochineel.* In: Hans Sloan.
A VOYAGE TO THE ISLANDS
MADERA, BARBADOES,
NIEVES, ST. CHRISTOPHERS
AND JAMAICA. VOL. II.
London, 1707–[1725.]

44.
[Naseberry or Sapodilla.]
In: Nikolaus Joseph, Freiherr
von Jacquin. SELECTARUM
STIRPIUM AMERICANARUM
HISTORIA. Vienna, 1763.

45.
*A capivard or Water hog at
the foot of a bananier.* In:
François Froger. RELATION
OF A VOYAGE MADE IN
1695, 1696, 1697. London,
1698.

46.
[Maize.]
In: Giovanni Battista
Ramusio. TERZO VOLUME
DELLE NAVIGATIONI ET
VIAGGI. Venice, 1606.

47.
[Buffalo.]
In: Francisco López de
Gómara. PRIMERA Y
SEGUNDA PARTE DELA
HISTORIA GENERAL DE LAS
INDIAS. Zaragoza, 1553.

48.
[Armadilho.] and *[Pece
Muger.]*
In: Francesco Redi.
EXPERIMENTA CIRCA RES
DIVERSAS NATURALES.
Amsterdam, 1675.

49.
António [Pereira.] "[NORTH
AND SOUTH AMERICA.]"
[Portugal, ca. 1545.]

50.
*Ala espada y el compas. Mas.
y mas. y mas. y mas.*
In: Bernardo de Vargas
Machuca. MILICIA Y
DESCRIPCION DE LAS
INDIAS. Madrid, 1599

42c.

42e.

47.

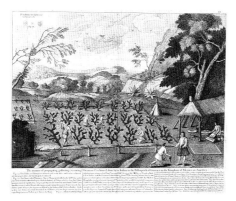

43.

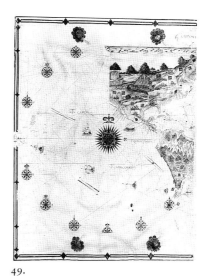

48.

44.

45.

49.

46.

50.

51.
[Map of the world.]
In: Battista Agnese.
["PORTOLAN ATLAS."]
[Venice, 1543–1545.]

52.
[Titlepage.] In: Wilhelm Piso.
HISTORIA NATURALIS
BRASILIAE. Amsterdam,
1648.

53.
[Landscape.]
In: Matthaeus Merian.
GRANDS VOYAGES, PT. XIV
(GERMAN). Hanau, 1630.

54.
[Cannibal scenes.]
In: Theodor De Bry. GRANDS
VOYAGES, PT. III (GERMAN).
Frankfurt, 1593.

55.
*Portrait du Combat
Entre les Sauvages.*
In: Jean de Léry. HISTOIRE
D'UN VOYAGE FAICT EN LA
TERRE DU BRESIL.
Geneva, 1580.

56.
[Weeping ceremony.]
In: Jean de Léry. HISTOIRE
D'UN VOYAGE FAIT EN LA
TERRE DU BRESIL.
[La Rochelle], 1578.

57.
Francois Carypyra. and
Louis Marie.
In: Claude d'Abbeville.
HISTOIRE DE LA MISSION
DES PERES CAPUCINS EN
L'ISLE DE MARAGNAN.
Paris, 1614.

58.
Jodocus Hondius. AMERICA
Amsterdam, [1619.]

Item 59 has been deleted.

59a.
[Alligator and Snake.] In:
Maria Sybilla Merian. OVER
DE VOORTTEELING EN
WONDERBAERLYKE
VERANDERINGIN DER
SURINAEMSCHE INSECTEN.
Amsterdam, 1719.

51.

57.

56.

52.

55.

53.

58.

54.

59a.

59b. *[Pineapple.]*
In: Maria Sybilla Merian.
DISSERTATIO DE
GENERATIONE ET
METAMORPHOSIBUS
INSECTORUM
SURINAMENSIUM.
Amsterdam, 1719.

60.
[Giovanni Battista Ramusio.]
LA CARTA UNIVERSALE
DELLA TERRA FIRMA &
ISOLE DELLA INDIE
OCCIDENTALI. Venice, 1534.

61.
[Atahualpa at Cajamarca.]
In: Francisco de Xérez.
VERDADERA RELACION DE
LA CONQUISTA DEL PERU.
Seville, 1534.

62.
*Il Cuscho Citta Principale
Della Provincia del Peru.* In:
Giovanni Battista Ramusio.
TERZO VOLUME DELLE
NAVIGATIONI ET VIAGGI.
Venice, 1565.

63.
Americaner in Peru.
In: Hermann Fabronius. DER
NEWEN SUMMARISCHEN
WELT-HISTORIE.
Schmalkalden, 1612.

64.
The Riche Mines of Potosí.
In: Augustín de Zárate. THE
DISCOVERIE AND CONQUEST
OF THE PROVINCES OF
PERU. London, 1581.

65.
[Titlepage.]
In: Theodor De Bry. GRANDS
VOYAGES, PT. VI. (LATIN).
Frankfurt, 1596.

66.
*Indi Hispanis aurum
sitientibus.*
In: Theodor De Bry. GRANDS
VOYAGES, PT. IV. (LATIN).
Frankfurt, 1594.

Item 67 has been deleted.

68.
Von jher Hirschjacht.
In: Theodor De Bry. GRANDS
VOYAGES, PT. II. (GERMAN).
Frankfurt, 1603.

69.
Oppidum Secota.
In: Theodor De Bry. GRANDS
VOYAGES, PT. I. (LATIN).
Frankfurt, 1590.

Item 70 has been deleted.

59b.

63.

65.

64.

60.

66.

61.

62.

68.

69.

71.
Comme les Dames de Dasamonguepeuc Portent Leurs Enfans.
In: Theodor De Bry. GRANDS VOYAGES, PT. I. (FRENCH). Frankfurt, 1590.

72.
Sumendi Cibum Modus.
In: Theodor De Bry. GRANDS VOYAGES, PT. I. (LATIN). Frankfurt, 1590.

Item 73 has been deleted.

Item 74 has been deleted.

Item 75 has been deleted.

76.
NOVA BELGICA ET ANGLIA NOVA. [Amsterdam, 1635.]

77.
Vitis Laciniatis Foliis.
In: Jacques Philippe Cornut. CANADENSIUM PLANTARUM. Paris, 1635.

78.
Of Potatoes of Virginia. In: John Gerard. THE HERBALL OR GENERALL HISTORIE OF PLANTES. London, 1597.

79.
Modo che tengono i medici nel medicare gl'ingermi.
In: Girolamo Benzoni. LA HISTORIA DEL MONDO NUOVO. Venice, 1565.

80.
Nicotiana.
In: Petrus Peña and Matthias de L'Obel. STIRPIUM ADVERSARIA NOVA. London, 1570 [i.e., 1571.]

81.
El Sassafras. In: Nicolás Monardes. PRIMERA Y SEGUNDA Y TERCERA PARTES DE LA HISTORIA MEDICINAL. Seville, 1574.

82.
[Martin Pring at Plymouth Harbor.]
In: Pieter Van der Aa. NAAUKEURIGE VERSAMELING. Vol. 22. Leiden, 1707.

83.
[Amerindians in Newfoundland.]
In: Johann Ludwig Gottfried, ed. NEWE WELT UND AMERIKANISCHE HISTORIEN. Frankfurt, 1631.

71.

72.

79.

80.

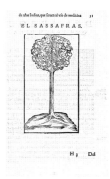

81.

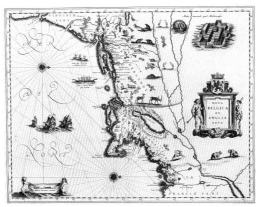

76.

82.

77.

78.

83.

84.
[Hunting in the New World.]
In: Johann Theodor De Bry.
GRANDS VOYAGES, PT. X.
(GERMAN). Oppenheim,
1618.

85.
William Blake. *Europe Supported by Africa & America.* In: John Stedman.
NARRATIVE, OF A FIVE
YEARS' EXPEDITION
AGAINST THE REVOLTED
NEGROES OF SURINAM. Vol.
II. London, 1796.

86.
Indianer können der Spanier.
In: Theodor De Bry. GRANDS
VOYAGES, PT. IV. (GERMAN).
Frankfurt, 1594.

87.
[Processing Sugar cane.]
In: Johann Ludwig Gottfried.
NEWE WELT UND
AMERICANISCHE
HISTORIEN. Frankfurt, 1655.

Item 88 has been deleted.

89.
[*"Newfoundland and Surrounding Area."* London, ca. 1671.]
From ["THE BLATHWAYT
ATLAS."] [London, ca. 1683.]

90.
Herman Moll. TO THE RIGHT
HONOURABLE JOHN LORD
SOMMERS . . . THIS MAP OF
NORTH-AMERICA. [London,
ca. 1730.]

91.
Virginia.
In: John Smith. A MAP OF
VIRGINIA. Oxford, 1612.

92.
This figure representeth the engine, to wind off the silk from the cods.
In: [Edward Williams.]
VIRGINIA'S DISCOVERY OF
SILKE-WORMES. London,
1650.

93.
Virginia Farrer.
A mapp of Virginia discovered to ye hills.
[London], 1651. From: ["THE
BLATHWAYT ATLAS."]
[London, 1683.]

84.

85.

89.

90.

86.

91.

92.

93.

94.
America.
In: Arnoldus Montanus. DE
NIEUWE EN ONBEKENDE
WEERELD. Amsterdam, 1671.

95.
*T'Fort Nieuw Amsterdam Op
De Manhatans.*
In: BESCHRIJVINGHE VAN
VIRGINIA, NIEUW
NEDERLANDT. Amsterdam,
1651.

96.
*Novae Sveciae seu
Pensylvaniae in America
Descriptio.*
In: Thomas Campanius
Holm. KORT BESKRIFNING
OM PROVINCIEN NYA
SWERIGE UTI AMERICA.
Stockholm, 1702.

97.
Herman Moll. A NEW &
EXACT MAP OF THE
DOMINIONS OF THE KING
OF GREAT BRITAIN.
[London, ca. 1730.]

Item 98 has been deleted.

99.
*Paysage de la Province de
Bemarin au Royaume
d'Apalache.* In: Charles de
Rochefort. HISTOIRE
NATURELLE ET MORALE DES
ILES ANTILLES DE
L'AMERIQUE. Rotterdam,
1681.

100.
Wie die Wilden in Florida.
In: Theodor De Bry. GRANDS
VOYAGES, PT. II. (GERMAN).
Frankfurt, 1591.

101.
*Primogeniti Solennibus
ceremonijs Regi sacrificantur.*
In: Theodor De Bry. GRANDS
VOYAGES. PT. II. (LATIN).
Frankfurt, [1609.]

102.
*La Terra De Hochelaga Nella
Nova Francia.*
In: Giovanni Battista
Ramusio. TERZO VOLUME
DELLE NAVIGATIONI ET
VIAGGI. Venice, 1556.

103.
*Le Grand Sacrifice des
Canadiens.*
In: CEREMONIES ET
COUTUMES RELIGIEUSES
DES PEUPLES IDOLATRES.
Amsterdam, 1723.

94.

95.

100.

96.

101.

97.

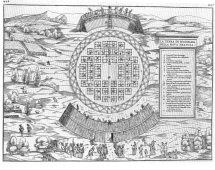

102.

99.

103.

104.
[Cannibal America.]
In: Ferdinando Gorges.
AMERICA PAINTED TO THE
LIFE. London, 1659.

105.
*[Emblem — Come over and
help us.]*
In: Increase Mather. A BRIEF
HISTORY OF THE WARR
WITH THE INDIANS IN NEW
ENGLAND. Boston, 1676.

106.
Et leges et sceptra terit.
In: Louis Armand de Lom
d'Arce, baron de Lahontan.
NEW VOYAGES TO NORTH-
AMERICA, VOL. 2. London,
1703.

107.
George Romney. JOSEPH
TAYADANEEGA CALLED THE
BRANT, THE GREAT CAPTAIN
OF THE SIX NATIONS.
[London], 1779.

108.
Canadiens en Raquette.
In: M. Bacqueville de la
Potherie. HISTOIRE DE
L'AMERIQUE
SEPTENTRIONALE.
Paris, 1722.

109.
*[French Showing Christian
images to the Indians.]* In:
Louis Armand de Lom
d'Arce, baron de Lahontan.
MEMOIRES OF DE
L'AMERIQUE
SEPTENTRIONALE, VOL. II.
Amsterdam, 1728.

110.
*Preciosa Mors Quorundam
Patrum é Societa. Jesu in
Nova Francia.*
In: François Du Creux.
HISTORIAE CANADENSIS.
Paris, 1664.

111.
*Plan of an American New
Cleared Farm.*
In: Patrick Campbell.
TRAVELS IN THE INTERIOR
INHABITED PART OF NORTH
AMERICA. Edinburgh, 1793.

112.
*The Figure of the Indians'
Fort or Palizado in New
England.*
In: John Underhill. NEWES
FROM AMERICA. London,
1638.

Item 113 has been deleted.

104.

109.

105.

106.

110.

107.

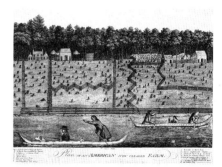

111.

108.

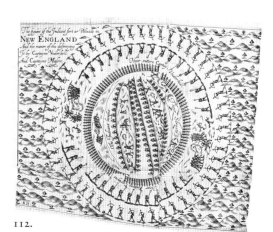

112.

114.
Paul Revere, engraver. *Philip. King of Mt. Hope.*
In: Thomas Church. THE ENTERTAINING HISTORY OF KING PHILIP'S WAR. 2d ed. Boston, 1772.

115.
John Foster. *A Map of New England.*
From: William Hubbard. A NARRATIVE OF THE TROUBLES WITH THE INDIANS IN NEW ENGLAND. Boston, 1677.

Item 116 has been deleted.

117.
Maurice Mathews, John Love, Thomas Nairn, Edward Crisp. A COMPLEAT DESCRIPTION OF THE PROVINCE OF CAROLINA IN 3 PARTS. [London, 1711.]

118.
[Buccaneer smoking a pipe.]
In: Alexandre Olivier Exquemelin. HISTOIRE DES AVANTURIERS. Paris, 1686.

119.
Rock Brasiliano.
In: Alexandre Olivier Exquemelin. BUCANIERS OF AMERICA. London, 1684.

120.
Sr. Hen: Morgan.
In: Alexandre Olivier Exquemelin. BUCANIERS OF AMERICA. 2d. ed. London, 1684.

121.
FRANCISCUS DRAECK. [Germany? ca. 1588.]

122.
Baptista Boazio. THE FAMOUSE WEST INDIAN VOYADGE MADE BY THE ENGLISHE FLEET. [London, 1589.]

123.
Guaura. In: William Hack. "AN ACCURATE DESCRIPTION OF ALL THE HARBOURS . . . IN THE SOUTH SEA OF AMERICA." [London, ca. 1698.]

124.
Pieter Pieterszoon Hein. BESCHREIBUNG VON EROBERUNG DER SPANISCHEN SILBER FLOTTA. Amsterdam, 1628.

114.

119.

120.

115.

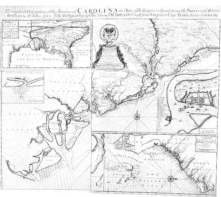

117.

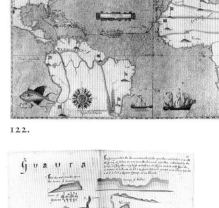

121.

122.

118.

123.

124.

125.
Caljou de Lima.
In: Joris van Spilbergen.
OOST ENDE WEST-INDISCHE
SPIEGEL DER NIEUWE
NAVIGATIEN. Leyden, 1619.

126.
[Fort Caroline.]
In: COPPIE D'UNE LETTRE
VENANT DE LA FLORIDE.
Paris, 1565.

127.
Saint Augustine. In: Walter
Bigges. A SUMMARIE AND
TRUE DISCOURSE OF SIR
FRANCES DRAKES WEST
INDIAN VOYAGE. London,
1589.

128.
Nicolas Visscher. NOVI
BELGII NOVAEQUE ANGLIAE
NEC NON PARTIS VIRGINIAE
TABULA. [Amsterdam: Petrus
Schenk, Jr., ca. 1715.]

129.
Reinier and Joshua Ottens.
TOTIUS NEOBELGII NOVA ET
ACCURATISSIMA TABULA.
Amsterdam, [between 1726
and 1750.]

130.
*"Long Island sirvaide by
Robartte Ryder."*
[London, ca. 1679.] From:
["THE BLATHWAYT ATLAS."]
[London, 1683.]

131.
[Louis Joliet.] "NOUVELLE
DÉCOUVERTE DE PLUSIEURS
NATIONS DANS LA
NOUVELLE FRANCE." [New
France, ca. 1674.]

132.
Cyprian Southack. CHART TO
SHEW YE ENGLISH THAT
LIVE IN THE PLANTATIONS
OF NORTH AMERICA.
Boston, 1717.

133.
*A Plan Representing the
Form of Settling the Districts
or County Divisions in the
Margravate of Azilia.* In:
Robert Mountgomery. A
DISCOURSE CONCERNING
THE DESIGN'D
ESTABLISHMENT OF A NEW
COLONY TO THE SOUTH OF
CAROLINA. London, 1717.

125.

126.

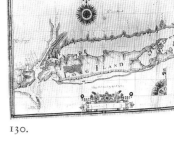

130.

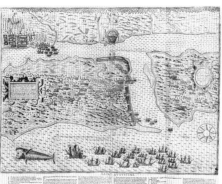

127.

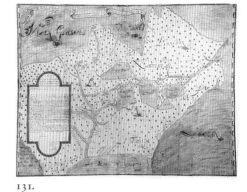

131.

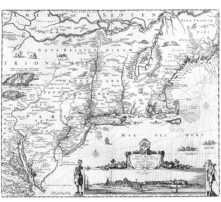

128.

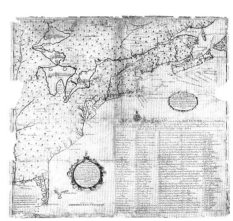

132.

129.

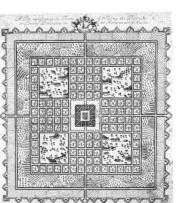

133.

134.
Captain Ince. A VIEW OF
LOUISBURG IN NORTH
AMERICA. [London], 1762.

135.
[Moses Harris.] A MAP OF
THE SOUTH PART OF NOVA
SCOTIA. [London], 1750.

136.
Thomas Johnston. QUEBEC,
THE CAPITAL OF NEW
FRANCE. [Boston, 1759.]

137.
S.G., Sieur de Longchamp.
CARTE DES POSSESSIONS
FRANÇOISES ET ANGLOISES.
Paris, 1756.

138.
Thomas Kitchin. A NEW AND
ACCURATE MAP OF THE
BRITISH DOMINIONS IN
AMERICA. [London, ca.
1764.]

139.
John Mitchell. A MAP OF THE
BRITISH AND FRENCH
DOMINIONS IN NORTH
AMERICA. [London], 1755.

Item 140 has been deleted.

134.

137.

135.

136.

138.

139.

Item 141 has been deleted.

142.
John Verelst. SA GA YEATH QUA PIETH TOW KING OF THE MAQUAS. [London, 1710.]

143.
John Verelst. ETOW OH KOAM KING OF THE RIVER NATION. [London, 1710.]

144.
John Verelst. HO NEE YEATH TAW NO ROW KING OF THE GENERETHGARICH. [London, 1710.]

145.
John Verelst. TEE YEE NEEN HO GA ROW, EMPEROR OF THE SIX NATIONS. [London, 1710.]

146.
Wiliam Verelst. TOMO CHACHI MICO OR KING OF YAMACRAW, AND TOOANAHOWI HIS NEPHEW. [London, ca. 1735.]

147.
THE BRAVE OLD HENDRICK THE GREAT SACHEM OR CHIEF OF THE MOWHAWK INDIANS. [London, ca. 1740.]

148.
Samuel Blodget. A PROSPECTIVE PLAN OF THE BATTLE NEAR LAKE GEORGE. Boston, 1755.

149.
Thomas Pownall. A DESIGN TO REPRESENT THE BEGINNING AND COMPLETION OF AN AMERICAN SETTLEMENT OR FARM. London, 1761.

150.
Peter Gordon. A VIEW OF SAVANAH. [London, 1734.]

142.

146.

147.

143.

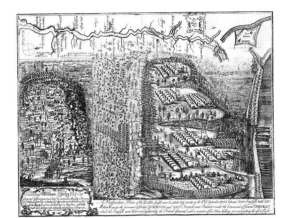

148.

144.

149.

145.

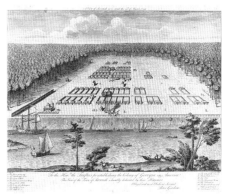

150.

151.
THE ICHNOGRAPHY OF
CHARLESTOWN AT HIGH
WATER. [London], 1739.

152.
[John Carwitham.] AN EAST
PERSPECTIVE VIEW OF THE
CITY OF PHILADELPHIA.
London, [ca. 1774.]

153.
Nicholas Scull and George
Heap. A MAP OF
PHILADELPHIA, AND PARTS
ADJACENT. WITH A
PERSPECTIVE VIEW OF THE
STATE-HOUSE. [Philadelphia,
1752.]

154.
Thomas Pownall. A VIEW OF
BETHLEM. London, 1761.

155.
John Carwitham. A VIEW OF
FORT GEORGE WITH THE
CITY OF NEW YORK FROM
THE S.W. London, [after
1764.]

156.
Thomas Howdell. A SOUTH
WEST VIEW OF THE CITY OF
NEW YORK. [London, 1763–
65.]

157.
Thomas Howdell. A SOUTH
EAST VIEW OF THE CITY OF
NEW YORK. [London, 1763–
65.]

158.
John Carwitham. A SOUTH-
EAST VIEW OF THE CITY OF
BOSTON IN NORTH
AMERICA.
London, [ca. 1794.]

159.
Earlom, THOMAS POWNALL
ESQR. London, 1777.

160.
Thomas Pownall. A VIEW OF
THE GREAT COHOES FALLS,
ON THE MOHAWK RIVER.
London, 1761.

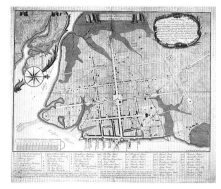

151.

152.

153.

154.

155.

156.

157.

158.

159.

160.

161.
Thomas Pownall. A VIEW OF
THE FALLS ON THE
PASSAICK. London, 1761.

162.
Thomas Pownall. A VIEW IN
HUDSON'S RIVER OF
PAKEPSY & THE CATTS-KILL
MOUNTAINS. London, 1761.

163.
Thomas Pownall. A VIEW IN
HUDSON'S RIVER OF THE
ENTRANCE OF WHAT IS
CALLED THE TOPAN SEA.
London, 1761.

Item 164 has been deleted.

Item 165 has been deleted.

165a.
[George Ehret.] *Magnolia.*
In: Mark Catesby. THE
NATURAL HISTORY OF
CAROLINA, FLORIDA, AND
THE BAHAMA ISLANDS. Vol.
II. London, 1743.

165b.
Phoenicopterus Bahamensis.
In: Mark Catesby. THE
NATURAL HISTORY OF
CAROLINA, FLORIDA, AND
THE BAHAMA ISLANDS. Vol.
I. London, 1731.

166.
THIS SR. IS THE MEANING OF
THE QUEBEC ACT. [London],
1774.

167.
THE BOSTONIANS IN
DISTRESS. London, 1774.

168.
THE BOSTONIAN'S PAYING
THE EXCISEMAN, OR
TARRING & FEATHERING.
London, 1774.

169.
THE PATRIOTICK BARBER OF
NEW YORK. London, 1775.

170.
J. DeCosta. A PLAN OF THE
TOWN AND HARBOUR OF
BOSTON. London, 1775.

161.

167.

162.

163.

168.

165a.

165b.

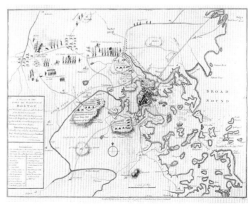

169.

166.

170.

171.
Bernard Romans. AN EXACT
VIEW OF THE LATE BATTLE
AT CHARLESTOWN JUNE
17TH, 1775. [Philadelphia,
1775.]

172.
Balthazar Frederic Leizelt.
VÜE DE LA NOUVELLE
YORCK. Augsburg, [ca. 1776.]

173.
Balthazar Frederic Leizelt.
VÜE DE PHILADELPHIE.
Augsburg, [ca. 1776.]

174.
LA GRANDE BRETAGNE
MUTILÉ. Amsterdam, [ca.
1765.]

175.
DEDIÉ AUX MILORDS
D'AMIRAUTÉ ANGLAISE PAR
UN MEMBRE DU CONGRÉS
AMERICAIN. [Augsburg?]
1778.

176.
...L'ETAT DE LA NATION
D'ANGLETERRE. Amsterdam,
[ca. 1780.]

177.
THE WISE MEN OF GOTHAM
AND THEIR GOOSE.
[London], 1776.

178.
THE BOTCHING TAYLOR.
[London], 1779.

179.
POOR OLD ENGLAND.
[London], 1777.

171.

177.

172.

173.

174.

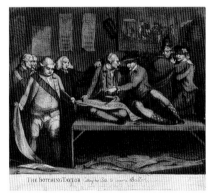

178.

175.

179.

176.

180.
[AMERICA TOE] HER
[MISS]TAKEN [MOTH]ER.
[London], 1778.

181.
[BRITANNIA TOE]
AMER[EYE]CA. [London],
1778.

182.
BRITANIA AND HER
DAUGHTER. A SONG.
[London], 1780.

183.
THE TEA-TAX TEMPEST, OR
THE ANGLO-AMERICAN
REVOLUTION. [Nuremberg,
1779.]

184.
LABOUR IN VAIN. [London],
1782.

185.
JOHN BULL TRIUMPHANT.
London, 1780.

186.
Sebastian Bauman. PLAN OF
THE INVESTMENT OF YORK
AND GLOUCESTER.
[Philadelphia, 1782.]

187.
REDDITION DE L'ARMÉE
ANGLOISES. Paris, [ca. 1781.]

188.
THE RECONCILIATION
BETWEEN BRITANIA AND
HER DAUGHTER AMERICA.
[London, 1782.]

180.

181.

185.

182.

186.

183.

187.

184.

188.

189.
Borel. L'AMÉRIQUE
INDÉPENDANTE. Paris, 1778.

190.
Alexander Robertson.
MOUNT VERNON IN
VIRGINIA. London, 1800.

191.
Amos Doolittle. A DISPLAY
OF THE UNITED STATES OF
AMERICA. New Haven,
[1788.]

192.
An American log-house.
In: Victor Collot. A JOURNEY
IN NORTH AMERICA. Paris,
1826.

193.
*View of a saw mill & block
house upon Fort Anne Creek.*
In: Thomas Anburey.
TRAVELS THROUGH THE
INTERIOR PARTS OF
AMERICA. London, 1791.

194.
American stage wagon.
In: Isaac Weld Jr. TRAVELS
THROUGH THE STATES OF
NORTH AMERICA. 4th
edition. London, 1800.

195.
[Andrew Ellicot.]
PLAN OF THE CITY OF
WASHINGTON IN TERRITORY
OF COLUMBIA. Philadelphia,
1792.

196.
Abraham Bradley. MAP OF
THE UNITED STATES,
EXHIBITING THE POST
ROADS. [Philadelphia], 1796.

197.
Osgood Carleton. THE
UNITED STATES OF
AMERICA. Boston, [1791.]

189.

190.

191.

195.

192.

193.

194.

196.

197.

198.
Preparation for War to defend Commerce; Pennsylvania Hospital, in Pine Street; Back of the State House; The Water Works, in Centre Square; An Unfinished House, in Chesnut Street; and Second Street North from Market St. wth. Christ Church. In: William Birch. THE CITY OF PHILADELPHIA. Philadelphia, 1800.

Item 199 has been deleted.

200.
AN EMBLEM OF AMERICA. London, 1800.

198.

200.

ENCOUNTERING
THE NEW WORLD
1493 TO 1800

designed by Gilbert Associates
was printed by Meridian Printing
on Gleneagle paper.

The typeface is Sabon,
designed by Jan Tschichold in 1966.
It is a variation of the Garamond style
and was named after Jacob Sabon,
a punchcutter from Lyon
who is thought to have brought
some of Garamond's matrices
to Frankfurt.

The book was bound by The Riverside Group.
1250 softcover and 250 casebound copies
for the John Carter Brown Library

December 1991